Urban Youth in China: Modernity, the Internet and the Self

Routledge Research in Information Technology and Society

1. Reinventing Government in the Information Age
International Practice in IT-Enabled Public Sector Reform
Edited by Richard Heeks

2. Information Technology in Government
Britain and America
Helen Margetts

3. Information Society Studies
Alistair S. Duff

4. National Electronic Government
Building an Institutional Framework for Joined Up Government—A Comparative Study
Edited by Martin Eifert and Jan Ole Püschel

5. Local Electronic Government
A Comparative Study
Edited by Helmut Drüke

6. National Governments and Control of the Internet
A Digital Challenge
Giampiero Giacomello

7. The Politics of Cyberconflict
Security, Ethnoreligious and Sociopolitical Conflicts
Athina Karatzogianni

8. Internet and Society
Social Theory in the Information Age
Christian Fuchs

9. Hacking Capitalism
The Free and Open Source Software Movement
Johan Söderberg

10. Urban Youth in China: Modernity, the Internet and the Self
Fengshu Liu

Urban Youth in China: Modernity, the Internet and the Self

Fengshu Liu

Routledge
Taylor & Francis Group
New York London

First published 2011
by Routledge
711 Third Avenue, New York, NY 10017

Simultaneously published in the UK
by Routledge
2 Park Square, Milton Park, Abingdon, Oxon OX14 4RN

Routledge is an imprint of the Taylor & Francis Group, an informa business

First issued in paperback 2013

Typeset in Sabon by IBT Global.

Library of Congress Cataloging-in-Publication Data

Liu, Fengshu.
Urban youth in China : modernity, the internet and the self / by Fengshu Liu.
p. cm. — (Routledge research in information technology and society ; 10)
Includes bibliographical references and index.
1. Internet—China. 2. Youth—China. 3. Social change—China. I. Title.
HN740.Z9I5676 2011
305.2350951'091732—dc22
2010029038

ISBN13: 978-0-415-87486-1 (hbk)
ISBN13: 978-0-415-85191-6 (pbk)

To my father Liu Fang, who passed away in 2008 and the thought of whom has accompanied me throughout this project. To my mother, Zhang Guiying, who has always been there for me.

Contents

Acknowledgments

This book has its origin largely in a seminar organized by the research group TransAction at the Institute of Educational Research, University of Oslo in 2009. It is at that seminar that Julian Sefton-Green, upon listening to my presentation of a paper, enthusiastically encouraged me to write a book on Chinese youth and the Internet. Julian has helped me throughout the process of producing this book by giving both intellectual and practical assistance, for which I am deeply grateful. My gratitude goes equally to Ola Erstad, the leader of TransAction, whose full support I have been enjoying over the past years.

I gratefully acknowledge the financial and institutional support by three institutions. The book is an outcome of a three-year (2007-2009) project funded by the Norwegian Research Council (NFR) and hosted by the Institute of Educational Research (PFI) at the University of Oslo, where I have been enjoying strong support from many of the academic staff and administrative personnel in the past years. The institute extended my postdoctoral term to facilitate the completion of the book, for which I am very grateful to the head of the institute, Ola Stafseng. I am also deeply indebted to the Norwegian Non-Fiction Writers and Translators Association (NFF) for awarding me the 'Book Project' stipend for this book.

I give special thanks to all the research participants and my personal contacts during the fieldtrips for their strong support in various forms, without which this book would not have been possible.

The book has benefitted from the comments of four peer readers: Berit Karseth, Jon Lauglo, Terfot Ngwana and Andrew Dobelstein. I am fortunate to have known Berit, who has been keeping me pleasant company on my scholarly journey ever since she became my PhD supervisor in 2003. As usual, her critical and yet 'sympathetic' reading of my text has been greatly helpful. Berit has an unusual ability to spot the 'gems' in my writing and encourage me to crystallize them, from which I have benefited in reflecting upon the Chinese 'dual modernity'. Jon is another colleague who has shown great support for this book from the beginning. He has patiently commented on the larger issues of the book as well as helped me say more with fewer words. Terfot has been a patient discussion partner. I thank him very much for perusing the manuscript and for stimulating my interest in the Chinese 'dual' self. Andrew drew on his experience and knowledge in political science and sociology as well as his personal experiences in China to make comments on my draft.

I would also like to extend my thanks to the three peer reviewers of the book proposal I sent to Routledge. They warmly supported the project and offered useful comments.

I thank the journal *New Media and Society* for permission to adapt text for certain sections in chapter 5 from the following article: The Norm of the 'Good' Netizen and the Construction of the Proper Wired Self: The Case of Chinese Urban Youth. *New Media and Society*. XX(X) 1–16. 2009. Chapter 6 draws heavily on: It Is Not Merely about 'Life on the Screen': Urban Chinese Youth and the Internet Café. *Journal of Youth Studies*. 12 (2): 167–184. (2009). Chapter 4 is partly based on text adapted from: The Internet in the Everyday Lifeworld: A Comparison between High School Students in China and Norway. *Comparative Education*. 46 (4): 527–50. (2010)

I have enjoyed enthusiastic support from people at Routledge from the very beginning. A special thank to Benjamin Holtzman, who received my project with great warmth and interest. My thanks also go to Jennifer Morrow and Max Novick. I am also thankful to Eleanor Chan, Jasmine Perez (the copyeditor), and Marianna Vertullo (the copyediting manager) at IBT Global.

I am deeply grateful to my family in China and Norway for their unfailing love and support, not least for their patience with my absence and 'absent-mindedness'.

October, 2010, Oslo.

Introduction

Three decades may seem to be a long time in an individual's life. However, it hardly counts as a moment in human history. And yet, in just such a moment of history, dramatic changes can take place, transforming millions of people's lives. China serves as a vivid illustration of such a social drama. Since the end of the 1970s, when China began reforms and opened up to the world with its pronounced goals of modernization, the world has witnessed a range of rapid, profound and interrelated changes in that society. These include, *inter alia*, the shift from a planned to a market economy, fast economic growth, rapid popularization of the new information and communication technologies (ICTs), increased socio-geographical mobility, fragmentation and individualization of society and the emergence of a whole generation of only-children (due to the one-child policy first implemented in 1979). Accompanying these changes is China's increased interaction with the world, which has been intensified by its inauguration into the World Trade Organization (WTO) in 2001 and its hosting of the 2008 Olympic Games. The social and cultural forces generated by these changes contend with the continuing authoritarian rule by the one-Party regime and with the rejuvenation of some traditional values in the reform era, making the Chinese conundrum all the more complicated. Where this huge country is heading and what implications its transformation (in any form) may have for both Chinese society and the rest of the world have aroused much public debate over the past three decades.

There has been a sense of urgency and a strong interest in understanding Chinese society around the world, especially in the West, where China is often perceived as a newly emerging superpower and a threat to the existing world order. Studies on reform-era China[1] have yielded a substantial body of work from a range of disciplines such as political science, international relations studies, economics, cultural studies, sociology, history and the humanities. The introduction of the Internet to China has further reinforced the world's interest in understanding how Chinese society may evolve, or even be revolutionized.

Yet, despite the considerable research effort devoted to understanding present-day Chinese society, there is a striking dearth of research on Chinese youth, the core group of individuals who will play a pivotal role in China's transition and its future social, cultural, economic and political activities.

The tendency to leave youth out of the discussion about social change is also reflected in research on the Chinese Internet. According to the China Internet Network Information Center (CNNIC, 2008),[2] with 253 million netizens (Internet users) by the end of June 2008 (in comparison with 22.5 million in 2001), China surpassed the U.S., which was originally the country with the largest number of netizens in the world. The number of netizens in China reached 384 million by the end of 2009 (CNNIC, 2010). However, despite phenomenal growth, the Chinese Internet has so far been a predominantly urban-youth phenomenon, with 10–30-year-olds, mostly members of the only-child generation, as the main group of netizens. And yet, existing research has scarcely focused on this group, in spite of the significant amount of research generated by Chinese Internet studies. My review of literature on the Internet in China, in line with the meta-review conducted earlier by Kluver and Yang (2005), indicates that the greatest amount of academic attention on this topic has been paid to two areas. One is the political implications of the new medium for China, one of the last 'communist' countries with its authoritarian state apparatus, hence the governmental control of the Internet. In this school of writing, the view from outside the country remains dominant (Bucher, 2004; Damm, 2007; Zhou, 2006). The other is the development of China's telecommunications and Internet infrastructure, the studies of which are mostly in the form of large-scale surveys done by business-oriented agencies and the China Internet Network Information Center (CNNIC), China's officially assigned Internet research and administrative agency. This latter strand of research mostly focuses on the quantitative aspects, like users and usage data, without offering insights into the reasons and intentions that lie behind these data. Thus, although existing research has dealt with important issues concerning the Chinese Internet, there seems to be a lack of interest in 'micro'-level analysis of the experiences of individual users. It largely remains a question how the Internet is interacting with China's social and cultural fabric (not merely the political system) and how in the process of being appropriated by young people, it is both being shaped and contributing to the shaping of the Chinese society.

Given that both youth and the Internet hold the potential to inflict, or at least contribute to, far-reaching social change, it is important to address this knowledge gap by systematically investigating how youth are interacting with the Internet in a Chinese context. This is the rationale behind this book. It will explore how urban youth, members of the only-child generation, appropriate the Internet in negotiating meaning making within the socio-cultural context of a Chinese society experiencing spectacular transformation. My investigation will be organized around two general themes: young people's perceptions and experiences of the Internet and their online self-representation. In so doing, it is hoped that this book will contribute to the world's understanding of Chinese society by shedding light on what it means to be Chinese today and how 'modern' 'Chineseness' may be (re) constructed in the Internet age.

THEORETICAL POINT OF DEPARTURE

My theoretical point of departure is that just as the Internet needs to be understood within its specific socio-cultural context, the ways in which people interact with the Internet are closely interrelated with their everyday life actualities, that is, the biographical (for example, their only-child status in the case of the young people in this book), social, cultural, historical and institutional context in which their lives are embedded. My perspective thus is to stress the interrelatedness between users' online and offline lives, an approach which is sensitive to the socio-cultural embeddedness of young people's experiences of the Internet and their identity work on the Internet.

Obvious as it may appear, such a perspective is often downplayed in existing research on young people and the Internet (Buckingham, 2008; Lee, 2005; Liu, 2010). Much research on this subject has been influenced by the notion of the 'Internet Generation' (alternatively, the Net generation, the digital generation, the e-generation and the wired generation) (e.g., Prensky, 2001, 2006; Tapscott, 1998), the generation raised steeped or immersed in digital technology, especially the Internet, with great implications for education and future society (Oblinger and Oblinger, 2005; Tapscott, 1999). Although useful for designating the Internet's far-reaching impact on today's young people, the notion of the 'Net Generation' often adopts a fairly standardized perspective from within the developed Western world, assuming that the Internet bears the same meaning for young people (with access) all over the world. There is little recognition of the relationship between who they are and the socio-cultural and biographical situations in which their lives are embedded. Rather, access to the Internet and age are the only determinants of the 'Net generation'. As such, it tends to 'treat young people as a monolithic social entity and see their relationship to technology as sequential and unidirectional, rather than complex and co-constructed' within specific local contexts (Lee, 2005: 316; see also Buckingham, 2008). This also reflects a tendency in much of the existing Internet research to be concerned with the Internet primarily through such notions as 'cyberspace', or 'virtuality' (Miller and Slater, 2000; Slater, 2001; Hine, 2000). By treating the Internet as 'a kind of placeless place', *apart from* the rest of social life (Miller and Slater, 2000: 4), this approach tends to downplay the interrelatedness between the user's offline and online life. Furthermore, by treating the Internet as a globally uniform culture in its own right, transcending local and national boundaries, it sheds little light on locally specific Internet experiences.

Thus, despite the large amount of Internet research, we still lack knowledge about how young people, and indeed, people in general, in specific local, socio-cultural contexts are appropriating the Internet in ways that make sense for their own purposes. This gap can only be addressed through a close examination of the lived experiences of the Net by individuals (and groups) as active agents living in contemporary societies. The qualitative studies conducted by Miller and Slater (2000) and Bakardjieva (2005) represent some useful attempts in this direction. Miller and Slater (2002), in their ethnographic study of the Internet in Trinidad, have documented not merely the diverse meanings and applications that the Internet finds at the hands of users in Trinidad, but the social, political and cultural contexts which account for the local people's Internet experience. Bakardjieva (2005) offers an ethnographic account of how 'ordinary' Canadians interpret, domesticate and creatively appropriate the Internet by integrating it into the projects and activities of their everyday lives. A common contribution of the two studies is a detailed account of the complex integration between online and offline worlds based upon rich qualitative data, showing that people's Internet-related meaning making and experiences are embedded in their everyday life actualities.

Viewed in this way, young people's perceptions and experiences of the Internet reflect their experiences as real persons in their offline socio-cultural environments (Giese, 2006; Rheingold, 2000; Turkle, 1995). This means that in order to gain a better understanding of the relationship between young people and the Internet, rather than studying them as mere Internet users, we need to take a holistic approach to them as human beings, taking into account who they are (or who they perceive themselves to be) in the 'real' life. Or in Schutz's (Schutz and Luckmann, 1973) words, we need to locate them in their 'everyday lifeworlds' and 'social-biographical situations' with their specific 'relevance structures'. For this end, we need to view the Internet as a technology in interaction with some key institutions which affect young people's lives in significant ways, such as the state, the family, the school, higher education institutions, the market and the workplace.

It is noteworthy that recognizing the embeddedness of online activities in offline realities does not mean to say that online experiences do not affect offline realities. Online experiences may also influence people's perceptions of reality and behavior in physical places—at least potentially (Fletcher, 1998). Furthermore, recognizing the 'situatedness' of individuals' perceptions and experiences of the Internet does not mean that subjects are deprived of agency by their social conditions. Rather, just as it is wrong to render people as objects totally manipulable by technologies, as suggested by the crude technology determinism, it is equally wrong to assume that they lack action in the social context in which they find themselves. In other words, people are active, creative and expressive—albeit socio-culturally situated—subjects (Mackay and Gillespie, 1992: 698). That is, they actively appropriate technology, but this is done without stepping out of their everyday lifeworlds into a different 'finite province of meaning' (Schutz and Luckmann, 1973: 23).

The general theoretical perspective briefly outlined here will be developed more fully as the book unfolds. In particular, in each of the empirical chapters, I will provide a more specific theoretical lens in line with this more general perspective to view the specific questions I will explore there.

Seen through such a perspective, by focusing on Chinese youth and the Internet, this volume serves as an exploration of youth's more general experiences of China's reform and opening up under the auspice of 'socialism with Chinese characteristics' or what I shall call 'dual modernity'—a concept which I first use in Chapter 1 and to which I shall also return at the very end of the book. It is hoped that it will shed light on Chinese youth's identity negotiation and the kind of self emerging from this process.

THE CONCEPT OF YOUTH FROM A CHINESE PERSPECTIVE

The concept of youth varies from culture to culture and it is subject to reconstruction over time (Jones and Wallace, 1992). The equivalent term of 'youth' in Chinese is '*qing nian*', literally translated as the 'green years'. Despite commonalities with the Western notion of youth, the Chinese concept of '*qing nian*' differs from the former in some important ways. Kwong (1994) identifies two major differences in this respect, which although based upon her observation some years ago, seem still largely true. First, the Chinese notion of youth adopts a broader age range, with the cohort stretching beyond the narrow age band of what constitutes youth in most academic discussion in the West (one exception is Nilan and Feixa, 2006) to include people ranging from those in their early teens to those in their late twenties. Sometimes people below 35 (or even 38) years old are all referred to as '*qing nian*'. Secondly, in the West, at least historically, the word 'youth' refers to a relatively indeterminate state between childhood and adulthood, associated with inexperience, impulsiveness, resistance, deviance and rebelliousness, and so is often viewed as dangerous and disturbing (Wulff, 1995). In contrast, the Chinese term carries much more positive connotations. 'Youth', or '*qing nian*', is associated with hope, courage and dynamism. Chairman Mao once famously said while addressing Chinese youth, 'you are like the sun between eight and nine o'clock in the morning. The world is yours as well as ours, but it will eventually be yours'. This remark is still often quoted by people talking about youth in today's China.

Moreover, and related to the positive notion of '*qing nian*', youth as a group is seen by Chinese Communist Party (CCP) leaders since Mao as a political force that they can draw on to strengthen their ruling power and to achieve various national goals (Ngai et al., 2001). Therefore, the concept of youth in contemporary China is much concerned with the ideological aims of the CCP as well as the collective interest of the state. Youth played an important role in revolutionary movements such as the civil war leading to the establishment of the People's Republic of China (PRC) and the Cultural Revolution (1966–1976). Cultivating youth in line with the CCP's ideology and social-political goals has occupied a prominent position on the state's agenda in both Maoist and post-Mao China. The CCP after 1949 has maintained a large and extensive Communist Youth League (CYL). To demonstrate that youth work has always topped the agenda of the Party, in 2000 the Central Communist Youth League (CCYL) published a collection of remarks by top Chinese leaders Mao Zedong, Deng Xiaoping and Jiang Zemin, recognized as the core of three generations of Chinese leaders, regarding youth and youth work (*People's Daily*, August 2, 2000). Chinese official media lauded the publication of the book as a big event in the political life of youth all over China. The CCYL required that all Youth Leagues and all youth must study this book carefully and form a heated wave of studying it. This approach to youth is further reflected in the ideal youth culture package the Party leaders worked out for youth consumption in the post-Mao era, as I shall show in both Chapters 1 and 3. Thus, despite social transformation, the Chinese notion of youth has persisted in its association with the ideological aims of the state.

Using the broader notion of youth mentioned previously, China's youth hit 560 million in 2005 (Yan, 2006), exceeding the entire population of most countries in the world. Whereas Chinese youth share common characteristics, they are not a homogeneous group. Socio-geographical origins, different family backgrounds and other social and cultural factors generate heterogeneous outlooks just as in other societies. The most striking differences in the Chinese case are related to rural-urban origins. It seems that official statistics in China do not usually distinguish between rural and urban youth in talking about Chinese youth and thus far, most scholarly and media attention on Chinese youth has focused on the urbanites, especially college students (Yan, 2006). This tends to neglect the fact that the two groups can hardly be talked about in the same terms due to the rural-urban disparities largely as a result of the residency permits system known as *hukou,* strictly practiced under Mao and still valid in the post-Mao era, albeit to a lesser extent. CCP's differential treatment of rural and urban people puts rural Chinese, the majority of the population, at the bottom of the social hierarchy.[3] Rural people's second-class status compared with urbanites in China has been a well-known reality that has affected many generations and it still shapes today's rural young people's lived experiences of China's reform and opening up, as I shall show in more detail while discussing social stratification in present-day China in Chapter 1. This is reflected, among other things, by the fact that rural youth constitute the majority of the technological have-nots concerning access and availability of ICTs. As Yan (2006) rightly observes, today's rural youth, like their counterparts in the cities, are exposed to market values and competition and they tend to have similar life aspirations to those held by their urban counterparts. And yet, they are faced with much greater socio-cultural barriers that make it much harder for them to realize these aspirations than for urban youth.

The obvious disparities between rural and urban lives underlie urbanites' collective consciousness of their urbanite status just as it does rural people's self-perception as second-class citizens. As Fong (2007) observes, even the disadvantaged segments (e.g., those with laid-off parents, low incomes and poor academic performances) of the urban youth population tend not to think about themselves as part of a lower class. Rather, they see themselves, and are seen by society at large, as belonging to 'a superior urban citizenship category defined by its opposition to an inferior rural citizenship category' (Fong, 2007: 87). It is this segment of Chinese youth population, estimated at about 90 million (*China Daily*, January 20, 2007), the minority compared with the much larger number of their rural counterparts, that constitutes the most attractive part of the Chinese consumer market. They are members of the only-child generation and their families form the major component of China's emerging middle class (Anagnost, 2004).

In this book, I shall focus on urban youth, not because I wish to follow the trend to prioritize urban youth in academic and media coverage, but because they are the main group of China's netizens, as indicated previously, which in itself is a manifestation of the rural-urban disparities in China. There was little Internet access for people in the rural areas within my reach when I conducted this project, which reflects the general picture of Internet penetration in the country at the time.

A METHODOLOGICAL NOTE

The empirical findings presented in this book are an outcome of a three-year project on urban youth and the Internet in China. The primary aim is to obtain an in-depth understanding of urban youth's perceptions and experiences of the Internet and their online self-representation. For this purpose, it adopts a qualitative methodology, as the advantage of qualitative research is that it enables the researcher to capture the complexity and depth of perceptions and experiences of the research subject. Such a study will contribute to the 'micro'-level analysis of youth's experiences of *their* Internet in the context of present-day China, which is still scarce in research on the Chinese Internet.

Data collection was organized around four major topics: urban youth's I-and-the-Internet narratives; urban youth and the Net café, '*wangba*' in Chinese; youth's online community building; and youth's negotiation of a political self using the Internet. These are described next.

The first topic draws on semi-structured interviews with 70 young people, aged 15–28, during two fieldtrips in 2007 and 2008 in a city in North

China. The participants were recruited by a snowballing method, starting with my personal contacts. I conducted group interviews of 2–5 friends with some of the participants and individual interviews with the others depending on their preferences. The group interviews were conducted at a local café, where we could sit and chat in a relaxed atmosphere. I conducted the individual interviews at my home in the same city. The interviews lasted 1–2.5 hours.

The exploration of urban youth and the Internet café draws on observation at an Internet café and semi-structured interviews with 30 Internet café patrons. During June–September 2007, I dropped in at a number of Internet cafés in the earlier-mentioned city and settled down at one for more systematic investigation. I went to the café in the morning and afternoon nearly every day for three weeks. I also stayed there during a few lunch breaks. My stay varied between 1.5–3 hours. I recruited 30 young people (aged 15–24) for in-depth interviews (1–1.5 hours) to further explore from their perspective what *wangba* going is all about. Again, depending upon the participants' preferences, the interviews were conducted either individually or in groups (of 2–4 friends).

My investigation of the third topic, namely youth's online community building, is based on online observation at www.80end.com, a website constructed by and for China's post-80ers (born 1980–1989), *balinghou* in Chinese, and on content analysis of message threads collected during the observation. From April 2008 to January 2009, I visited the website once or twice a week to read the message threads, both the latest and the archived ones, and to observe what was going on in the forum. I paid special attention to the website's bulletin board system (BBS) (although I also looked into other sections of the website), as it has become the largest interaction area for the online community. It thus serves as a platform where the members can voice their perceptions and experiences of being post-80ers. Many of the message threads in the BBS shed light on the questions and issues related to their collective identity. The archived postings in the BBS constituted a large corpus of text. I chose to do a textual analysis of message threads (and the follow-up comments by the BBS members) containing opinions and topics that are relevant to the post-80ers' self-understanding and self-representation. These were downloaded and saved for analysis. The selected message threads amounted to about 160 in number. In conducting the content analysis, I adopted the process of coding described by Miles and Huberman (1994). This means that I classified the texts thematically, searching for 'coherent' meaning structures in the text material and paying attention to emergent concepts. In this process, words, phrases, sentences, paragraphs or whole message threads that belong to the same theme or topic are clustered together and given a label as a theme.

The topic of Chinese youth's negotiation of a political self using the Internet draws on semi-structured interviews with 38 Chinese university students (20 men and 18 women) in 2008. They were recruited using a snowballing method, starting with three of my former students, who were at a Chinese comprehensive university at the time. They passed on my research intention to some of their classmates, friends and acquaintances at both their own and other universities. These informants were from a range of disciplines at several universities. Twenty of these interviews were conducted on the phone from my office in Norway. The other 18 were done face-to-face in the earlier-mentioned city during my fieldtrip in 2008. All these interviews were conducted individually with the participants and they lasted 1–1.5 hours.

Besides these formal ways of data collection, I have, on numerous occasions, informally discussed with my relatives and friends in China the topic of young people and the Internet either on the phone, on Skype or face-to-face. This is a topic that seems to attract wide interest or concern in the Chinese urban community. In addition, the periods during which I stayed with relatives and friends in China in the past years provided me with good opportunities to do informal observations of (young) people's interaction with the new media. These all constitute valuable additional information that helps me to understand youth and their relationship with the Internet in a specific Chinese context. I make reference to these in this book when pertinent.

As one can see, the data for this project were collected both 'offline' and 'online' involving three main methods: interviews, observation and online content analysis. Altogether, the project has involved 138 participants, 10 months' online observation and three weeks' offline observation.

The 'offline' fieldwork was conducted in a city in North China during two fieldtrips, respectively in the summer of 2007 and autumn of 2008, which amounted to five months in the field. The city is a medium-sized (in Chinese terms) one with about 430,000 people. Like other cities in the country, it has been undergoing a rapid process of modernization. However, compared with the larger, more prosperous places like Shanghai, Beijing and the cities in the fast-developing south, this is one of the less developed cities in the country. As such, compared with those cities, this one may appear to be more limiting in some ways such as social life as well as opportunities for education and professional development. I decided to focus on youth in such a city for two main reasons. First, most research on the Chinese Internet has been done in big cites or more developed areas, especially metropolitan centers like Beijing, Shanghai and Guangzhou. Therefore, youth in less developed areas such as the city in this study are left at the margin of research. Secondly, I used to work and live in the city for some years before leaving China. I still have close relatives, friends and acquaintances living there. Thus, I could capitalize on my social networks there in conducting the fieldwork, which is important for the success of data collection, hence the trustworthiness of the study, as I shall show next.

Trust between the researcher and the participant is crucial for the trustworthiness of social research. But how to gain that kind of trust can vary from one cultural context to another. The fact that researchers' access to the field needs to be negotiated in culturally specific ways in China has been verified by my personal experience doing fieldwork there. In China, social science research is a relatively new phenomenon, especially when it involves methods meant for capturing people's perceptions and lived experiences from individuals' own perspectives. This has largely to do with the fact that both in history when the feudal emperors ruled and during the Maoist regime, the major questions about human life and society had set answers provided by the authorities. Questioning of society, hence the authority, was deemed neither necessary nor welcomed by the rulers bent on social order. China's opening up since the end of the 1970s made it possible for the development of social science studies at a dramatically faster pace and larger scale than ever before. China scholars worldwide now have the possibilities to collect first-hand data in the country. Thus, whereas social science was formerly dominated by ideological studies mainly controlled by the Party and by quantitative methods—where there was empirical research at all—the fieldwork 'boom' in recent years has brought about a methodological 'revolution' in China studies, as was noted by the researchers discussing their experiences in doing fieldwork in China at the Nordic Workshop on Fieldwork Methodology and Practice in China organized by the Nordic Institute of Asian Studies (NIAS) in October 2003.[4]

Nevertheless, conducting social research in China is still rather different from, say, in a Western country. Due to the Party's strong concern about social control, as will be discussed in Chapter 1, and hence, political censorship, so far social and political spaces for free debates and dialogue are still limited, as is evidenced by the governmental control of the Internet that I am going to discuss in Chapter 2. Thus, there is often skepticism against researchers going to people and discussing various social issues. This seems especially true for foreign researchers and Chinese researchers coming back from the West, who might be suspected by the Party officials of mobilizing the masses for social movement against the government. I remember when I was collecting data for a study on basic education in China's rural areas[5], one of the leaders, an acquaintance of mine, whom I turned to for help with the access to some official documents at a county-level educational committee, asked me, 'Are you not a spy for some Western agency working against China?' He was half serious and half joking. But still I had to explain much before he willingly let me in. In such a climate, people are not used to being interviewed for social research purposes and many are unwilling to participate for fear that the investigation may cause political trouble for them. This explains why—although this may not be unique to China—it is often hard to conduct fieldwork without help from personal contacts.

The experience of the constraints incurred by this phenomenon seems to be a shared one by researchers conducting social research in China, as evidenced by the discussion at the earlier-mentioned NIAS workshop. Often, as some of these researchers reported, the researcher has to go through sophisticated *guanxi* (social connections, relationships or networks) before gaining access to the field. In order to establish, develop or at least smooth *guanxi* for the purpose of fieldwork, gift giving, befriending through banquets, socializing and Karaoke with the various 'gate keepers' are often necessary. Whereas this may prove to be especially frustrating for foreign researchers, it does not mean it is easy for native researchers like me to conduct fieldwork in China. If one has friends and relatives that can introduce the research project to potential participants, things will be much easier and the trustworthiness of the study will be enhanced.

Given this second consideration and my wish to explore how youth in less developed urban areas interact with the Internet, I decided to do the fieldwork in the aforementioned city, where I did fully enjoy the help of my relatives and friends. I contacted some of them soon after I got the grant for this project, asking them to help me recruit participants. As usual, they showed their whole-hearted support. The snowball started to roll and grew quickly when I got the first informants among my relatives' and friends' children, who introduced me to their friends and schoolmates. During the two fieldtrips, I also got in touch with a few secondary schoolteachers, who were my former classmates or students, to help organize some interviews with their students. My personal contacts in China also proved helpful with the interviews with the university students (both online and offline) as well as with my investigation of youth and the Internet café.

All the interviews were conducted in Chinese. They were recorded with permission from the informants and transcribed. Pseudonyms are used and information about the individuals' identity and their institutions is kept to the minimum, especially in Chapter 8, which is on young people's negotiation of a political self in an authoritarian political context. Concerning the online observation and content analysis, as the site is open to all to read without the need to register as a member, I see its content as in the public domain (Franklin, 2003). Nevertheless, for the sake of anonymity, the online (nick)names of the participants and the time of publication of the message threads quoted will not be given in the text. As the postings are all in Chinese, I have translated them in a way that would best capture the original tone in the passage.

ORGANIZATION OF THE BOOK

The book consists of three background chapters and five empirical chapters. In line with the theoretical perspective outlined previously, the purpose of the first three chapters is to develop in more detail the context alluded to earlier for understanding youth's interaction with the Internet by providing systematic background information about present-day Chinese society, the Chinese Internet and urban youth. The five empirical chapters are illustrative cases of the two general themes of this book. These chapters cover the four major topics mentioned earlier: youth's Internet-related perceptions and experiences (Chapters 4–5); urban youth and the Net café (Chapter 6); youth's online community building (Chapter 7) and youth's negotiation of a political self in the Internet age (Chapter 8).

Chapter 1 sets an overall backdrop to the whole book by describing and analyzing the social transformation, both dramatic changes and significant cultural continuities, in reform-era China with its 'dual modernity'. It discusses both social transformation in material terms and the changing subjectivities in the face of competing 'subjectification technologies'. This chapter will lay a foundation for the analysis of the Chinese Internet and urban youth respectively in Chapter 2 and 3.

My point of departure in Chapter 2 is that like any other forms of cultural tools, the Internet is unavoidably affected by the socio-cultural context in which it is embedded, which is in line with the book's general theoretical perspective. Whereas to some this seems to be a kind of by-now-conventional wisdom, it has strangely been often ignored in Western discussions about the Chinese Internet. Seen through such a socio-cultural perspective, it is not a surprise that the Chinese Internet has come to bear its Chinese characteristics. The spectacular development in infrastructure and rapid and continuing growth in the number of users have been taking place in the context of stringent governmental control of the Internet. Such a paradox reflects the dynamics and complexities of a Chinese society undergoing rapid and profound transformation, with increasing wealth and changing subjectivities. A distinctly Chinese Internet culture is emerging from users' active appropriation of the new medium in ways that reflect their offline everyday actualities.

Chapter 3 focuses on urban youth, the major group of China's netizens. Here I provide a systematic portrayal of who today's Chinese urban youth are by giving an account of their generational characteristics, including the material conditions of their lives, their value orientations and self-understanding and their struggles and frustrations. On the whole, I show that it seems that Chinese urban youth, mostly members of the only-child generation, are characterized with a range of apparent paradoxes, or contradictions, which when put together, form a collective character that reflects the complexities of the specific socio-cultural context in which their lives are embedded. Faced with the expanding range of choices, but meanwhile limited by various constraints, they, as agents, are actively making their lifeworlds in interaction with others. This chapter and the preceding two chapters form the general background against which the empirical chapters will unfold.

Chapter 4 presents the major themes that have emerged out of the participants' 'I-and-the-Internet' narratives. Drawing on Alfred Schutz's notions of 'the everyday lifeworld', 'social-biographical situation' and 'relevance system', it explores how the Internet is actually being taken up by young people and made meaningful for themselves within the socio-cultural context of today's China. Assigning relevance and meaning of the Internet to their lives, the participants drew different aspects of their social-biographical situations into the foreground, such as, *inter alia*, only-child status, high parental expectations, the exam-oriented educational system, being young, lack of authority in the family and society, the competitive labor market and the 'emerging adulthood' among the only-child community.

Whereas Chapter 4 explores how Chinese urban youth relate to the Internet in ways that are meaningful to themselves, Chapter 5 examines how their relationship to the Internet is negotiated within a context of widespread Internet anxiety, which has been coproduced with larger social-political concerns in the reform era. This chapter takes up the tension between the competing discourses about 'proper' uses of the Internet in China, which reflects the disparities between the norms about the ideal subject form outlined by the 'subjectification regime' in reform-era China and young people's lived experiences of the Internet as a central symbol of modernity. It empirically explores how such a tension is being played out in the participants' Internet-related narratives.

It is hardly possible to talk about urban youth and the Internet without mentioning the Internet café, *wangba* in Chinese, given its popularity among Chinese urban youth—despite the derogatory connotations the *wangba* carries, official control and widely diffused access at home. Thus, in Chapter 6, I explore from the Internet café goers' perspective what *wangba* going is all about based on observation at *wangbas* and in-depth interviews with urban youth. Thus, this chapter continues to explore urban youth's relationship with the Internet, but extends the investigation to youth and the Internet café as a special social space. It shows how youth's perceptions of the café may be different from official and parental 'demonization' of the *wangba*.

Chapter 7 examines the collective identity construction of China's post 80ers in their virtual locality www.80end.com in the light of theorizing of online community building and collective identity construction. Based on content analysis of postings at the BBS of the website, where I did observation for ten months, I show that emerging from the many message threads I have analyzed is a particular collective self that not only counters the stereotype of the post-80s generation, but also goes far beyond a mere deconstruction or subversion. Their collective narrative revolves around three main themes: the 'uncollapsed we', the 'unique/multifaceted we' and the 'nostalgic we', which together represent their self-perception that defies and rejects any forms of simplified understanding of this generation. These discourses identified in their collective narrative are as much about who they are as who they aspire to be, that is, it is both about 'being' and 'becoming', both of which are essential for their negotiation of a particular collective self.

Chapter 8 explores why political indifference and nationalist passion coincide in Chinese cyberspace—an understudied subject. Although both are well-known phenomena by now, they are rarely examined side by side and as a continuum of the changing subjectivities in present-day China. Drawing on in-depth interviews with university students and examining the Chinese political self across seemingly contradictory online behaviors, this chapter argues that online nationalism constitutes an important extension of the political being in the Chinese context. In other words, one needs to put the political indifference and the nationalist passion together in order to gain an understanding of how Chinese youth negotiate a Chinese self in the context of rapid and profound social transformation due to various 'glocal' forces.

The book's concluding chapter brings together the major themes that have emerged from the preceding chapters to shed light on the interplay between the personal, the social and the new medium in shaping Chinese urban youth's identity construction. At the end, I shall reflect on the emerging Chinese self in the light of the concept of 'dual modernity'. I shall note that it is in keeping with Chinese social thought to make apparent contradictions appear to be complementary elements of a harmonious whole.

1 Social Transformation in China (1979–2010)

Since 1979, when China initiated its reform and opening up, Chinese society has been undergoing rapid transformation that has amazed the world. The pace at which this huge and ancient nation is changing is unprecedented in its own history and rare in any other country's history. It is against this general backdrop that the lives of the current generation of youth, as well as China's technological landscape, are unfolding. Therefore, one cannot hope to gain an understanding of today's Chinese youth, the Chinese Internet and youth's interaction with the new medium without an understanding of the social transformation in China in the past three decades.

I start with a brief historical overview of China's reform and opening up, with a special reference to the subsequent reconstruction of socialist theory under the new historical circumstances. I then describe the spectacular economic growth and discuss the discomforting re-stratification of Chinese society as a concomitant process. The chapter also discusses the changing value systems and the fragmentation of Chinese society. Before I conclude this chapter, I take up another aspect of the socio-cultural transformation in today's China, namely, the new subjectification regime. This is crucial for an understanding of the competing discourses affecting young people's subjectivities and hence identity construction, both in general terms and with regards to their interaction with the Internet. This chapter demonstrates that while focusing upon change, it is important to recognize the continuity in cultural and political values and practices. It also highlights that social transformation in China, as in other societies, needs to be understood in both material terms and non-material terms, by which I mean the changing subjectivities and subjectification technologies.

DUAL MODERNITY: 'SOCIALISM WITH CHINESE CHARACTERISTICS'

'Dramatic', 'rapid', 'profound', 'miraculous', 'amazing', 'incredible' and 'spectacular' are among the most often used expressions to describe China's socio-cultural transformation in the past three decades. Such impressive social change would not have been possible without Deng Xiaoping's radical reform policy initiated at the end of the 1970s.

Mao's social movements, especially the 'Great Cultural Revolution' (1966–1976), cast China into an abyss of political chaos and left it at the verge of economic bankruptcy. It was within this context of desperateness and emergency that at the Third Plenum of the Communist Party's Eleventh Central Committee on December 22, 1978, Deng Xiaoping, the then Vice-Premier of the State Council, proposed a pragmatic approach to China's development by discarding Marxist-Leninist dogmatism and introducing market-driven economic policies. In so doing, Deng rejected not only Mao's utopian vision of the egalitarian society of the Great Leap Forward (1958–1960) and the unending class struggle of the Cultural Revolution, but also the Stalinist model of state control of the economy, collectivization of agriculture and emphasis on heavy industry that China had followed since the 1950s. Deng's pragmatic approach, which is epitomized in his famous maxim that no matter whether it is a white or black cat, as long as it can catch a mouse, it is a good cat, emphasizes economic development as the key to virtually all the problems that China has faced in contemporary times. Economic development and population control were defined as two vital and interdependent strategies for the realization of the goal of the reform—the Four Modernizations, namely, the modernization of the country's agriculture, industry, national defense and science and technology. 'It is glorious to get rich' became a household slogan under Deng, encouraging everyone to strive to accumulate material wealth through hard work and market competition, which forms a sharp contrast with Mao's era when the emphasis on spiritual reward overrode material concerns and the mere mention of money might be condemned as practicing capitalism.

The reform started with 'eradicating the big-pot-meal system' (*feichu daguofan*), the Chinese colloquial expression for de-collectivization. In the countryside, where the economic reform was first initiated, farm land was leased to farmers to motivate them for increased production. Before the reform, under nearly 30 years of collectivism, agriculture had been under heavy government regulation through the collective production brigade. Farmers had little autonomy to decide either what to grow or how to handle their products. Another well-known reform policy aimed at developing rural non-agricultural sectors by transferring some of the labor force from the agricultural-based activities to local community enterprises. In the 1980s, the rural non-agricultural sector constituted the most dynamic component of the Chinese economy (Ma, 2003). In the city, the de-collectivization, which started in the early 1980s, took the form of smashing the 'iron rice bowl' (*tie fanwaner*). The 'iron rice bowl' stands for the employment and personnel management system in the city in Maoist China, which meant that urbanites' livelihood would be guaranteed from the cradle to the grave by lifelong employment and associated welfare services. Efficiency was not a concept in the management of personnel during the years when the focus was on class struggle rather than on productivity and central control was deemed as of ultimate importance. However, with the reform, terms like 'cost-recovery', 'efficiency' and 'competition' became the keynotes. This demanded a decentralization of decision-making and personnel management in the formerly centrally planned economy. Top administrators in business and education were thus given the right for hiring and firing, an authority that had never been granted before (Ma, 2003). Moreover, as the private sector in business and industry developed, attempts were also made to transform state industries and enterprises by contracting the state firms to 'trusted individuals' in order to improve efficiency and productivity. With the newfound employer freedom in the recruitment and allocation of labor, Chinese urbanites began to face layoffs and unemployment at the same time as they started to enjoy more individual autonomy in the market economy.

The economic reform also entailed a shift in emphasis from heavy industry to light and service industry. This adjustment soon proved to be vital for improving the overall living standards at a national level. People familiar with the picture of China during the Cultural Revolution may still remember that gray, dark blue and military green were the dominant colors of people's clothes, which indicated a plain and simple life. Lack of supply was common in those years. People had to use coupons for everyday consumption such as grain, oil, sugar, cotton, cloth, pork and even soap and detergent up till the early 1980s. But since the mid-1980s, coupons were gradually phased out and by the early 1990s they were no longer used (Ma, 2003). Today, the vast varieties of goods available in the market are not only for people's basic needs but also for the ever-growing desires for consumer choices within the context of rampant consumerism, as I shall discuss in further detail later on in this chapter.

As an integral part of the economic reform, China's opening up to the outside world began after the Third Plenum in 1978, when opening the door to external economic forces became official government policy in order to attract foreign investment. China used to trade extensively over the Silk Road and sea routes back in the Ming Dynasty (1368–1644). The Qing Dynasty (1644–1912), the one following Ming, however, banned trade with foreign countries until China was forced open by European imperialist powers during the Opium War (1839–1860). Under Mao, China largely walled itself off from the world and fell increasingly behind Europe, the U.S., Japan and some other Asian economies such as Korea, Singapore and Taiwan. But beginning in 1979, Deng put an end to China's decades of isolation from the rest of the world by starting with a number of Special Economic Zones (SEZs) located on the Southern coast. The SEZs offered foreign companies special incentives to invest, such as tax concessions and favorable labor regulations (Stockman, 2000; Shirk, 2007). For a while, the suppression of the demonstrators at Tiananmen Square in 1989 dampened people's trust in the Deng regime, and the promise of a burgeoning Chinese market seemed to be fading. However, following Deng's call for a new wave of market experimentation during his so-called 'southern excursion' (*nanxun*) in 1992, there has been an upsurge in foreign investment, especially from overseas Chinese in Southeast Asia.

Deng's pragmatism also made it possible, and indeed, necessary, to reinterpret the theory of socialism under the new historical circumstances. Since the commencement of the reform and opening up, there have been deliberate attempts by the Chinese leadership and its technocrats to reconstruct socialist theory based upon their reflections on China's practice of socialism since the founding of the People's Republic of China in 1949. The concept of 'socialism with Chinese characteristics' was initially put forward by Deng Xiaoping at the 12th Congress of the Chinese Communist Party (CCP) in 1982. Deng argued that in the process of the country's modernization, 'Marxism should be fixed to reflect Chinese reality and socialism should be the ideology with Chinese characteristics' (Deng, 1983: 52). 'Socialism with Chinese characteristics' has since served as the theoretical basis and justification for China's political, social, economic and cultural practices in the reform age.

Zhang (2004) has summarized the three major components of 'socialism with Chinese characteristics'. These include a political exclusivism that is centered on the sole leadership of the Party, an economic pragmatism that promotes whatever means conducive to the economic growth which is regarded by the Chinese leadership as the basis for the 'performance legitimacy' and a cultural particularism that addresses the problems caused by a moral vacuum at home and Western cultural influences from abroad (Zhang, 2004: 290). The political exclusivism is the baseline in the formulation of the 'socialism with Chinese characteristics'. The unchallenged political power of the CCP is emphasized as the guarantee for maintaining social and political stability that is in turn understood as the basis for continuous economic growth. An often-used support for this argument by the leadership is that the political transition of the former Soviet bloc countries serves as a negative example that China ought to avoid. The economic pragmatism is reflected in that in contrast to the political domain where the CCP power is exclusive, in the economic realm, power is allowed to be shared between the center and localities, between government organizations and enterprises and even between Chinese firms and foreign firms. In this sense, it is obvious that the economic power has been increasingly decentralized from within and internationalized from without whereas the political power has remained virtually unchanged and still centralized. The cultural particularism refers to the Party's emphasis on the uniqueness of Chinese culture, with a special appreciation of traditional Chinese culture. Accordingly, the contemporary relevance of Confucianism has been greatly emphasized. Since the early 1980s, various research institutes and foundations for Confucian studies, and even schools for children to learn

Confucian classics, have been established under the auspices of the government (Zhang, 2004: 289–290). The cultural particularism can be seen as an extension of the political exclusivism; and the combination of the two is meant as a counter balance to the market liberalism associated with economic pragmatism, so as to maintain the social-political status quo. As I show later on in this chapter, the mobilization of 'traditional' culture for 'modern' purposes is reflected in China's new regime of subjectification.

The three decades of practice of 'socialism with Chinese characteristics' has proved to be an uneven road, with a series of phased reforms, lending support to Deng Xiaoping's description of China's reform as 'crossing the river by feeling for stones at each step' or in the former General Secretary Hu Yaobang's words, 'an experiment without precedent'. Nevertheless, the core elements of 'socialism with Chinese characteristics' have remained consistent, which is also evident in the notion of building a 'harmonious society' (*hexie shehui*) advocated by the current leadership under President Hu Jintao as its vision for leading China into the twenty-first century. The 6th Plenum of the 16th Central Committee of the Communist Party of China concluded on October 11, 2006 with the passage of a resolution to establish a 'harmonious society' by 2020. Hu describes 'harmonious society' as a 'scientific development concept' which shifts China's primary focus from a purely economic growth model to a more balanced, Confucian-style approach aimed at maintaining growth while addressing daunting social issues such as the wide gap between the rich and poor, widespread environmental degradation and government and corporate corruption (Geis and Holt, 2009). Despite the shift of the primary focus from economic development to an overall social development, the emphasis on the CCP's rule and the efforts in promoting Chinese traditional culture and mitigating foreign influences in a rising China are evident in Hu's socialist 'harmonious society'. Thus, post-Mao China remains authoritarian although it has continued to build on policies promoting openness and integration with the international community (Geis and Holt, 2009).

In this way, China has reinvented itself through the Party leaders' hybridization of two seemingly contradictory ideologies: economic liberalism and political authoritarianism. Such a dual approach has been a result of China's response, not least resistance, to a constantly changing world—and a changing China—brought about by two grand forces, modernization at home and globalization in the world (Zhang, 2004; K. Liu, 2004). Chinese leaders see modernity as both signifying hope and danger for China (Bakken, 2000). Modernization and globalization, it is believed, would not only bring challenges to China economically, but culturally as well. It is a mainstream understanding among the Party leaders that whereas China could keep a certain distance from foreign cultures in the past, there is now no such 'pure land' on which China can be immune from the contamination of them within the context of globalization and the information revolution. The strategy advocated by the leaders to cope with this cultural challenge is centrally reflected in the notion of 'two civilizations'—'socialist spiritual civilization construction' and 'socialist material civilization construction'—which is meant to both reap the economic benefits of the reform and opening up and avoid the real and potential harms or 'spiritual pollution' brought about by modernization and globalization.

China's reform and opening up can thus be summarized as 'dual modernity'. I use this concept to denote China's state policy and the related sociocultural conditions under 'socialism with Chinese characteristics'. As I shall discuss in further detail in a later section of this chapter, such a duality bears directly on governance, which takes the form of a double 'subjectification regime'.

The 'dualism' here represents a search for a modern identity for China in the age of globalization (Lin, 2006; K. Liu, 2004; Zhang, 2004). This process has entailed two apparent paradoxes: a search for 'continuity with, as well as a departure from, the socialist legacy and cultural tradition of the past, and integration with, as well as resistance to, global capitalism' (Yu, 2009: 6). The contradictions and complexity inherent in such a process have led commentators to describe present-day China in postmodernist terms. For example, Sheldon Lu (2000: 146) sees postmodernism as 'the emergent cultural logic in contemporary China', characterized by 'the superimposition of multiple temporalities; the pre-modern, the modern, and the postmodern, which coexist in the same space and at the same moment.' Similarly, Terry Eagleton (1997) suggests in his essay 'The Contradiction of Postmodernism' that postmodernism—which he defines as the culture of late advanced capitalism—can always be seen as both a conservative and a radical force. And this dual character of the postmodern, he argues, is perhaps truer of contemporary China than any other place. Many other observers who see the world through Western cultural lenses would agree with the comments mentioned here. However, it seems that China's 'dual modernity' is in keeping with traditional Chinese views of the constituent parts of harmony—as I shall elaborate at the very end of this book.

In short, 'socialism with Chinese characteristics', or what I call 'dual modernity', under the auspice of which China's reform and opening up has been unfolding, underlies the dramatic transformation of Chinese society. As I shall show in further details in the rest of the chapter, just as it has spurred fast economic growth at the price of sharpened social disparities, it has likewise entailed the changing subjectivities and the state's new subjectification regime.

THE ECONOMIC MIRACLE AND RE-STRATIFICATION OF CHINESE SOCIETY

In December 2008, China celebrated the thirtieth anniversary of the Third Plenum that marked the beginning of the country's era of reform and opening up. Chinese had good reasons to celebrate this event. Since the Third Plenum, one of the most eye-catching achievements in the country is the miraculous economic growth. The country has sustained average annual growth rates of around 10% in most of the years since the reform started and its high growth survived the Asian financial crisis of 1997–1998. In 2009, in the midst of the global financial crisis, its economy still grew by 8.7%, achieving the central government's full-year growth target of 8% (China Statistical Bureau, 2010). Data from the World Bank, the Asian Development Bank and the United Nations unanimously point to three decades of extraordinary increases in income and consumption, vast reductions in poverty rates in both rural and urban areas and improvements in adult literacy, infant mortality and adult life expectancy (Candelaria, Daly and Hale, 2009). The proportion of the population living in poverty in China fell from 65% in 1981 to 4% in 2007, during which time more than half a billion people were hoisted above the poverty line, according to the World Bank (2009).

One of the major factors that lie behind the development is China's open-door policy after its long periods of isolation. As a result of the government's active encouragement through various policy measures, foreign direct investment (FDI) in China has grown rapidly since 1978. By mid-1996 there were over 120,000 foreign companies and joint ventures operating in China, employing 17 million workers (You, 1998: 170). An isolated, autarchic economy thirty years ago, China has absorbed around half of the foreign direct investment going to developing countries since the early 1990s (Garnaut, 2008). Its volume of foreign trade increased twenty-five times between 1978 and 2001 (Wu, 2005). Chinese exports, many of them manufactured in factories partially owned by foreigners, streamed out to world markets, starting the 'made-in-China' era. China surpassed the United States in 2007, becoming the second largest exporter and is forecast by the World Trade Organization (WTO) to pass Germany in the next few years (WTO, 2008).

China's rapid economic development and increasing influence on the global stage since the end of the 1970s paved the way for its accession to the WTO in 2001 and hosting of the 2008 Olympic Games. China was the East Asian region's anchor of growth when financial crisis hit its neighbors in the late 1990s. Analysts turned their eyes to China again during the world financial crisis that started in 2008, wondering if China would be the anchor of global growth (Garnaut, 2008). In short, in just less than three decades, China went from being a closed society to one that is highly integrated with the international economy. The tremendous economic growth has made China one of the biggest and most dynamic economies in the world, reviving the country as a regional and world power after over a century of humiliating weakness (Guthrie, 2006; Shirk, 2007).

With the rapid economic growth, talk of the 'middle class' (*zhongchan jieji*) in China has already gone beyond media reports, moving into official documents (*People's Daily Online*, October 27, 2004). According to a report by the Chinese Academy of Social Sciences (CASS) (2004, cited in *People's Daily Online*, October 27, 2004), families with assets valued from 150,000 yuan (then, $18,137) to 300,000 yuan ($36,275)[1] can be classified as middle class, and they are the group of people who enjoy stable incomes, are capable of purchasing private residences and cars and can afford the costs of education and holidays. The report suggests that the 'middle class' accounted for 19% of the country's 1.3 billion population by 2003. It further points out that assuming 1% annual growth, the 'middle class' is expected to make up 40% of the total population by 2020 (*People's Daily Online*, October, 27, 2004). Although the number of China's 'middle class' may be greatly exaggerated (Wang, 2008), the overall economic success has generated newly emergent confidence and a national pride among Chinese people, which saw its climax during the 2008 Beijing Olympics. As shown by the *Pew Global Attitudes Project* (Pew Research Center, 2005), Chinese people are now the most optimistic in the world about their future and are confident that the growing opportunities they have experienced in recent years will continue to expand.

Nevertheless, there is another side of the coin. The stress on market-oriented economic growth without large efforts to reconcile it with social amenity has resulted in a highly stratified society, with the richest 10% holding 45% of the country's wealth, and the poorest 10% holding only 1.4% (*China Daily*, June 19, 2005). Since the early 1980s, China, by many common measures of income inequality, has been transformed from one of the most egalitarian to one of the most unequal societies in the world. China's Gini coefficient, the internationally accepted measure of a country's income inequality (0 corresponds to complete equality and 1 to a single person having all the income), was less than 0.3 when the reform started, lower than developing countries in other parts of the world (World Bank, 1997). However, the 1997 World Bank report showed that it grew from 0.288 in 1981 to 0.388 in 1995, and that the rate of increase was similar in urban and rural areas. More recently, it has reached somewhere between 0.46 and 0.49, approaching 0.50 according to various Chinese official sources, above the 'alarm level' of 0.4 as defined by the Chinese government (*China Daily*, June 19, 2005).

The fast growing income inequality in post-Mao China is partly due to the greater growth advantages enjoyed by coastal provinces over inland ones (Candelaria, Daly and Hale, 2009; Meng, Gregory and Wang, 2005; Fan, Kanbur and Zhang, 2009). Meanwhile, the inequality trends in China also reflect fundamental changes in the way labor is allocated and rewarded in post-Mao China. As the market reform deepened, a socialist planned economy with fixed wage scales and virtually no labor mobility was replaced by a market-oriented system featuring a dynamic non-state sector and an increasingly open labor market. Exploiting this newfound autonomy, many workers left state employment to 'jump into the sea' (*xiahai*)[2] in the hope to maximize their economic benefits (Giles, Park and Cai, 2006). Income inequality increased sharply as labor mobility across urban regions and between rural and urban areas increased and the rate of return to different levels of labor market skills widened (Meng, 2000; Riskin, Zhao and Li, 2001; Khan and Riskin, 2000; West and Zhao, 2000; Yang, 1999). Just as it allows many to improve their incomes—and in some cases become rich, this process has also led to substantial layoffs, retirements and exits from the labor force (Giles, Park and Cai, 2006).

Besides, inequality also has to do with the well-known fact that in post-Mao China fortunes are sometimes made not through hard work and ingenuity, but through official corruption. Or, in Shirk's (2007: 31) words, 'Wealth and political power are closely intertwined because the government still plays such an important role in economic life.' Moreover, *guanxi* (social connections, relationships or networks), which is often closely related to corruption, has proved to be another important asset in one's pursuit of benefits in China's market economy, hence the re-stratification of society. *Guanxi* helps those who occupy or have occupied positions of power to retain or extend their social advantages (Gold, 1985; Johnson, 1997; Kipnis, 1997; Stockman, 2000; White, 1996; Yan, 1996; Yang, 1994). A commentator interviewed by *China Daily* (February 18, 2005) rightly argued that income should not be the only index used to define the middle class in China. Other elements such as social status, profession and *guanxi* must also be taken into account (see also Guthrie, 2006).

China's inequality is not only found between individuals and households, but also between social groups. The most obvious divergence in incomes is between rural and urban people. According to Nolan and Paine (1987), the gap between urban and rural incomes, which had been rigidly maintained during the Mao era, was at first narrowed by the change in policy after 1978. Reforms were implemented more speedily in the rural economy and resulted in a rapid rise in rural incomes. However, this narrowing of the urban-rural gap was not sustained beyond the first few years after the reform. According to official statistics, average rural incomes were 42% of urban incomes in 1978, rose to 54% by 1983, but then fell back again to 41% by 1990 (Zhao, 1993). The gap in disposable income between the city and the countryside jumped from 2.51 : 1 in 1998 to 3.23 : 1 in 2003 (Wang, 2008). The annual growth rate of urban incomes (8–9%) is almost twice that of rural incomes (4–5%) (Shirk, 2007). The income gap between China's affluent city dwellers and its huge rural population expanded to its widest level ever in 2009 due to the economic slowdown during the first three quarters of the year, putting millions of rural migrants out of work. According to *The Straits Times* (January 16, 2009), city dwellers earned an average annual income of 15,800 yuan in 2008, whereas the average rural income was 4,700 yuan.

That is, the gap between urban and rural incomes became 11,100 yuan in 2008, compared with 9,646 yuan in 2007.

The rural-urban inequality is also reflected in the uneven distribution of consumer goods among the population. Consumer goods are mainly associated with the so-called 'middle class', which, as we can see from my earlier description, so far still constitutes the minority of the population despite its growth in recent years. Despite the increase in inequality also among urbanites, it is urban residents, mostly only-child families, that constitute the main body of China's emerging middle class (Anagnost, 2004), perpetuating the historical pattern of urban-rural disparities in the country, as indicated in the Introduction. China's rural residents buy only 39% of goods output although they are the majority of Chinese population (*People's Daily*, October 27, 2004). It is estimated that the consumption capacity of the rural population is falling behind urbanites by at least 15 years (Wang, 2008).

The apparent differences in living standards among individuals, between rural-urban lives and the disparity between less-developed cities and big metropolitan centers resemble the socio-geographical disparities found in the world. The extreme contrast between the lavish lifestyles of the rich and the hardships of the poor is evident everywhere in today's China just as it is at the global level. Shifting rapidly in structure and composition, the stratification in Chinese society has come to resemble a class system replacing the Maoist socialist rank order. The awareness of these differences—sharpened by today's intensified global cultural flows through various means, such as the media, tourism and migration—shapes people's imaginations of 'the good life' and underlies the rural-urban, inland-coastal and China-abroad migration in the reform era.

The inequality has become a well-known problem that threatens social and political stability in China. The recognition of this emerging social crisis underlies the ongoing campaign of 'western development' launched in 1999 aimed at pulling people in the western regions out of poverty and the more recent Chinese governmental initiative of constructing a 'harmonious society', as mentioned earlier in this chapter. Nevertheless, it remains to be seen whether, to what extent and when China can achieve this new social-political goal, given the fragmented nature of Chinese society, which is reinforced by the social restratification discussed in this section.

CHANGING VALUES AND THE FRAGMENTATION OF CHINESE SOCIETY

With the deepening of the market-oriented reform and opening up, Chinese society has seen far-reaching value changes, resulting in a weakening in the Communist hegemony and a social fragmentation. Lin's (1999: 19) description of this process, written slightly over a decade ago, still rings largely true today:

> During the masses' rush to set up businesses, to speculate on stocks and stamps, to build real estate empires, and to chase whatever is in fashion, society is experiencing a sharp decline in moral beliefs such as those associated with collectivism, altruism, patriotism, and an emphasis on spiritual reward rather than material gratification. Lei Feng, the soldier once established as the national role model and glorified for his selflessness and loyalty to the Party, has lost public appeal. The desire to be more individualistic and independent in thinking supersedes the government's appeal to maintain Lei Feng's role-model status.

The mass media, which have themselves undergone dramatic marketization and exposure to global forces, have reinforced this process. They have displayed tensions between diversification and commercialization, on the one hand, and communist domination on the other (Stockman, 2000; Yu, 2009). Notwithstanding this, since the late 1970s, there has been an enormous expansion and proliferation in all forms of media, such as newspapers, magazines, radios, televisions, telephones and the Internet. Although still bound by their political obligations to the Party state, all these compete for the audience's attention by means of attractiveness in both styles and contents. Satellite transmissions from Taiwan and Hong Kong are widely available. Various forms of foreign programs, such as films, serialized classics, soap operas, music and sports, constitute a considerable part of the output by the state-controlled channels themselves. Besides, a wide range of foreign media products such as CDs, CVDs and DVDs—often bootlegged—are within easy reach in the cities (Litzinger, 2002). Moreover, access to (broadband) Internet has become a reality in the majority of urban households and is entering more and more people's lives in the countryside. Internet cafés, which have become popular in the city, are the hot spots for young Chinese to congregate (Litzinger, 2002; Liu, 2009). Despite the strong censorship, the Internet opens up unprecedented spaces for information and communication, broadening people's horizons and providing spaces for self-expression and identity experimentation. In short, the fast-developing and profit-driven mass media have allowed for a more diverse and open network of communication in China over the past three decades. This has also been reinforced by other processes, such as tourism, migration and absorption of foreign investment.

This has made a greater diversity of interpretations available to the Chinese population, resulting in interrelated processes such as pluralization of value systems, individualization and fragmentation of society. Intersecting with all these processes is the widespread consumerism (see Davis, 2005). Indeed, one can hardly talk about the value changes in post-Mao China without mentioning the rampant consumerism. As Qiu (2003: 14) observes,

Chinese society 'has swapped, at an astonishing speed, from Maoist puritan lifestyle in the 1970s to periodic anti-materialism campaigns in the 1980s and then suddenly to the flourishing of hedonism at the turn of the century'. Or as K. Liu (2004: 82) put it, 'Western consumer culture, or popular culture, has now found its largest marketplace in the world's most populous country.'

Advertising, which had been rare in the Maoist era and which reemerged in the late 1970s, has become a flourishing industry. China is currently one of the world's largest advertising markets (Research and Market Report, 2007). Newer products advertised have had a great impact on consumer habits in China, where many people, especially youth, have strong interest in consumer goods and the advertising of them and show high levels of knowledge of global brand names and beliefs as to the relative superiority of brands (Wang, 2008).

In addition to the flourishing advertising industry, evidence of the victory of consumerism is everywhere in the city no matter whether one takes a glimpse at people's homes or the street. Electric goods, such as color televisions, telephones, CD-and-DVD players, refrigerators and, more recently, mobile phones and computers, have become part of household belongings of almost every urban family, and indeed, an increasing number of rural families as well. It has been observed by some researchers that in terms of ownership of information and communication technologies, affluent urban Chinese households are not dissimilar to those in developed countries (Lim, 2008). Chinese people, especially trend-conscious urbanites, update their electronic equipment regularly (Lim, 2008; Wang, 2008). Cars have also become part of the budgeting of more and more urban families as a growing number of people are joining the so-called 'middle class'. As the flourishing advertising industry and other media images of consumer culture increase and inspire new wants, there is heightened pressure to keep up with one's peers (Gamble, 2003). Moreover, as I shall show further in Chapter 3, the one-child policy has also reinforced consumerism by fostering a child-centered culture. Only-children play an important role in urban families' household consumer choices (Chan and McNeal, 2004).

Today, one indeed cannot walk the streets of any Chinese city without taking note of the range of commodities and services available in street stalls and fancy department stores, all in sharp contrast to the Maoist era. Star hotels, restaurants, karaoke, disco halls, beauty parlors, fashion shops, McDonald's, Kentucky Fried Chicken, sports stadiums (often with indoor swimming pools)—just to name some—have become common in the city. Moreover, cosmetics stores featuring luxury brands such as Gucci, Christian Dior and Chanel have appeared in more and more Chinese cities. Meanwhile, the city's visible landscape is changing daily as old streets and buildings are being relentlessly pulled down and new, modern ones are mushrooming up: skyscrapers, highways, towering concrete apartment buildings, star hotels and sprawling supermarkets and malls, just to name a few. People are looking forward to moving into their new modern apartments, stepping on the newly paved roads and strolling in the newly built parks and shopping centers while being plagued by the noises of city construction day and night.

Another window through which we can gain a glimpse of the prevalence of consumerism is the changing wedding customs of urban young people. According to *People's Daily Online* (June 3, 2003), as recently as the 1950s–1960s, with just a few sweets and a plate of peanuts, two rolls of bedding and revolutionary songs sung loudly, a wedding was done. Toward the 1970s, people began to pay some attention to wedding and dowry, which included the 'three turn-rounds and one bell-sounding', namely bicycle, electric fan, sewing machine and a radio receiver. However, this was furthered in the 1980s, with the 'three big ones' taking over: the TV set, refrigerator and washing machine. But now, what is required for the wedding, which in most cases combines Chinese traditional customs with Western ones, includes an apartment and a sedan car along with many other types of modern equipment. The same article (in 2003) refers to a survey conducted in the whole city of Tianjin showing that on the average a newly married couple would spend 191,100 yuan on their wedding, almost 50 times that in 1978. These expenses are mainly spent on housing, household electric appliances, furniture and a banquet at the wedding ceremony. Buying and furnishing the new house takes an average expense of 110,900 yuan. That used for buying electric appliances averages 20,100 yuan and the expense on banquet is 9,300 yuan on the average. Moreover, honeymoon tours, photographs taken in (mostly Western-style) wedding garb and so forth are common practices. According to the same article, most of today's urban young people are determined not to get married until they (together with their families) have accumulated the economic basis for a 'middle-class' life, which is confirmed by my own study of China's first-generation young adult only-children (Liu, 2006a).

In short, the Western middle-class lifestyle, made known to the Chinese masses via the media and other forms of global cultural flows, has become the image of 'the good life' for Chinese people (Liu, 2008a). As Wang (2008) observes, few would disagree that the rampant consumerism in China today is a mockery of the Communist revolution, during which such consumerist pursuit would have been nipped in the bud. With this change, the emphasis has shifted from plain living and collective welfare to private materialistic gains and comfort and individual lifestyle. Furthermore, this process partly underlies the 'depoliticization' of Chinese society, whereby generally speaking people are now much less concerned about politics than they were in the 1980s, when the reforms had just taken off (Damm, 2007; Rosen, 2004, 2009; Xu, 1999; Yan, 2006).

To sum up, fundamental value changes have taken place in post-Mao China, whereby society has undergone a far-reaching process of fragmentation and, related to this, individualization (Yan, 2009). As a result, China has transformed from a society imbued with Maoist collective values to one featured with individualism and consumerism as the new dominating value system. How, then, does the Party state attempt to govern its 'newly liberated' subjects under the new social conditions?

THE MAKING OF THE NEW CHINESE SUBJECT

Foucault (1997: 81) defines 'government' as 'techniques and procedures for directing human behavior' or 'the conduct of conduct' (Foucault, 1982: 220–221). In examining the 'conduct of conduct' in post-Mao China, I am providing what Rose (1996a, 1996b) refers to as a 'genealogy of subjectifiction' of reform-era China. According to Rose, writing from a Foucauldian tradition, a 'genealogy of subjectification' would be concerned with localized attempts to produce meaning, especially as this occurs through various authorities in relation to wider socio-political interests. Following Rose (1996a: 300), questions central to such a genealogy may include (but are not limited to): What aspects of human beings become the focus of concern, regulation, shaping or reformation? What techniques or means are employed to govern the human being? What subject form do they seek to produce? What images or exemplars of the ideal persons are promoted or assumed by these practices? And finally, what are the goals of such practices and how are they related to the wider socio-political interests of today's China?

As indicated previously, subjectification regimes are related to wider socio-political interests. This means that the 'conduct of conduct' may vary from society to society. In the case of China, it is closely related to the Party state's practice of 'dual modernity'. Contrary to many people's prediction, China's proclaimed modernization project, with the introduction of a market economy and opening up to the world, has not brought about democratic political reforms. Instead, inherent in the notion of 'socialism with Chinese characteristics' is an emphasis on social-political control, which is seen as the guarantee of the country's economic development. As shown by the expanding body of China studies informed by the theory of governmentality (e.g., Dutton, 1992; Hoffman, 2003, 2006a, 2006b; Sigley, 2004, 2006; Greenhalgh and Winckler, 2005), instead of a total 'retreat of the state', the 'reform and opening up' process has generated governmental responses that have involved 'a creative blending of neoliberal rationalities and revitalized forms of socialist rationalities' (Sigley, 2006: 504). This has resulted in the co-existence in today's China of a highly liberal economic system based upon a neoliberal logic of the free market and a highly authoritarian political system under a one-Party leadership whose political lines are still much based on socialist norms and ideals (Eagleton, 1997; Zhang, 2004).

Such a dualistic approach to modernity presupposes a double subjectification regime, with two sets of subjectification technologies, often with conflicting values. One enjoins the subject to be the self-enterprising individual—autonomous, rational and free-choosing. This requires the subject to put the self at the center, thus placing the authority in the hand of the individual subject in the construction of identity. The economic reform sent a clear message to the Chinese masses that one now can—and indeed, is supposed to—be the master of one's own life by choosing the 'right' things to do in the free market rather than living a largely 'pre-destined' and 'egalitarian' life in the communist work unit or the rural commune, as was the case under Mao. As Zhang (2008) notes, with the commencement of the reform and opening up, China witnessed a quick prevalence of the term *ziyou* (freedom), such as *geren ziyou* (individual freedom), *yanlun ziyou* (freedom of speech), *xiaofei ziyou* (freedom in consumption), *ziyou shichang* (free market) and *ziyou jingzheng* (free competition). This is markedly different from the situation in Mao's era and in most other periods in Chinese history, where 'individual freedom' was largely erased and people's lives were largely under the control of institutions (Hsu, 1948; Yan, 2009).

Indeed, the discourse of autonomy, free choice and self-enterprising has been producing new subjectivities in reform-era China, establishing new norms of behavior and relations to the self and transforming individuals into 'responsible' and 'self-reliant' neoliberal subjects (Hoffman, 2003, 2006a, 2006b; Liu, 2006a, 2008a; Zhang, 2008). This enables the government to govern 'at a distance' (Rose, 1996a: 61). In this sense, Gidden's theorization of identity construction in advanced liberal democracies under the so-called 'late-modern' conditions seems to fit the Chinese case as well. Like their counterparts in those societies, with the reform, Chinese people are also offered the 'freedom' to choose, or they 'have no choice but to choose how to be and how to act', to use Gidden's words (1994: 75). Research shows that the kind of 'neoliberal' 'self-choosing and self-enterprising' ethos has been widely internalized by Chinese people, especially the younger generation (Hoffman, 2003, 2006a, 2006b; Liu, 2006a, 2008a; Zhang, 2008).

However, this is only half of the story. At the same time as the 'neoliberal' norms of governing are shaping Chinese subjectivities, another set of subjectification technologies is being staged. This calls on people to be the loyal subjects to the authoritarian Party state, who enjoins the subject to 'deny' one's self and embrace the communist-collectivist values such as selflessness, collectivism, a hard and plain (*jianku pusu*) life and devotedness to the Party, the people and the motherland. In so doing, it presupposes the authority the Party state is supposed to continue to have over individual subjects in guiding their choice and identity construction (Liu, 2008a).

This latter set of subjectification technologies works under the auspices of the 'socialist spiritual civilization construction', which has deliberately been introduced by the Party leaders as a counter-force to the liberalization concomitant to China's reform and opening up, as mentioned earlier in this chapter. According to the Party leaders, market economy and opening up to the outside world have brought about widespread and increasing evils of individualism, loss of belief in communism, money worship, materialism, hedonism, gambling, prostitution and corruption—all indicating self-centeredness and general deterioration in public morality, or a 'moral slide', threatening the social order crucial for China's success in its 'socialist modernization' (Bakken, 2000). Socialist modernization, the Party leaders emphasize, depends on both 'material civilization' *and* 'spiritual civilization', respectively representing economic development and social stability. It is in an attempt to integrate the two sets of subjectification technologies that the social engineers have worked out the theory of 'socialism with Chinese characteristics', which might well have been called 'neoliberalism with Chinese characteristics' (Liu, 2008a). In the social engineers' understanding, it seems to be no problem, and indeed imperative, for one to combine the two sets of enjoinments. Thus, the ideal form of subjecthood, as proposed by the Chinese social engineers, is supposed to incorporate both the qualities of the autonomous neoliberal subject in the free market *and* the communist-collectivist values of a socialist subject combined with Confucian self-cultivation and 'traditional Chinese virtues', as I shall discuss in further detail next.

One cannot talk about the ideal subjecthood without mentioning the term '*suzhi'*, which has been roughly translated as 'quality' but the connotations of which go far beyond that. As has been noted by some researchers (e.g., Anagnost, 2004; Bakken, 2000), '*suzhi*' is hardly a neologism (for a genealogy of *suzhi*, see Bakken, 2000; Kipnis, 2006), but it acquired new discursive power when it became conjoined with the idea of population in the reform era based on the authority's problematization of human conduct as indicative of 'low quality' of the population. To foster a whole generation of high quality people, the leaders called for the birth of 'fewer and better-quality' children, which served as part of the justification for the one-child policy (Fong, 2004a, 2007; Greenhalgh and Winckler, 2005; Woronov, 2002). However, the cultivation of *suzhi* does not stop there. It emphasizes both the production of a healthy offspring and the after-birth fostering, cultivating and disciplining of human beings (Bakken, 2000). Moreover, neither is the *suzhi* discourse limited to the only-child generation. According to the social engineers, the quality of workers and people in general should be improved. And, what is more important, such quality is supposed to be all-rounded (*quanmian*), which concerns not only their cultural level of knowledge and skills necessary for the national economic development but also their ideological, political, moral and behavioral cultivation conducive to social order. As such, the discourse of human quality describes human technologies closely linked to both the material and the spiritual side of 'socialist modernization'

(Bakken, 2000), or the earlier-mentioned double subjectification regime. As I shall show later on in this volume, such a notion of the ideal personhood is also reflected in the norm about one's 'proper' relationship with the Internet, the central symbol of modernity.

Bakken (2000), in his book *The Exemplary Society*, offers an account of how the Chinese state's social engineering attempts to cultivate the ideal subject based on exemplary norms. He shows that, to this end, various 'scientist-rational' measures and programs have been developed by the social engineers, such as the construction of personality models (*renge moshi*), launching campaigns, using propagandas and model figures, assessment and evaluation, all of which is reminiscent of the 'biopower' Foucault (1990) has described. The so-called 'science of personality' (*renxue*) was established by some technocrats to investigate the question of the desirable human qualities in modern Chinese society based on the double-requirement (Bakken, 2000). A person with modern quality, it is contended, embodies values such as striving forward, risk taking, innovation, competition and independence. As Bakken (ibid.) notes, a streak of social Darwinism can be easily discerned running through the entire discourse on modern human qualities. Competition is defined as the striving for superiority and the ability to exist according to the rule of the 'survival of the fittest'. Meanwhile, however, it is maintained that although in modern knowledge-based society, a talent's quality should be geared mainly toward production, the moral and ideological quality of the talent is also important, reflecting the discourse of the 'two civilizations'. In this ideal personhood with the double requirement, one can hardly miss the continuity of the notion 'both red and expert' (*youhong youzhuan)* addressed to the successors to the revolutionary cause in Mao's era and the Chinese traditional ideal for a talent 'possessing both virtue and competence' (*decai jianbei*).

Whereas the former aspect of the ideal personhood appeals to the population's self-interest in the free market economy, the latter proves to be less natural in a society where one's well-being is left to one's own choice and responsibility and where consumerism supersedes interest in living a plain and hard life. Therefore, the Party leaders have found it necessary to deliberately emphasize the latter set of human qualities in guiding the emerging subjecthood. The ongoing moral education in the era of reform and opening up represents a significant attempt at guiding the ideal subjecthood through exemplary norms with a view to strengthening the latter set of human qualities. Chinese leaders believe that to successfully halt the dangers of modernity as is reflected in the aforementioned 'moral slide', one has to 'bind' individuals (ibid.: 103). Therefore, individual self-restraint and self-control have become key concepts in the moral education in reform-era China. This reflects the basic theme of Confucian ethics, the major goal of which is to socialize individuals to suppress personal desires and personal identity in social interactions (Hwang, 1999; Hsu, 1948). It is argued that self-control represents a conscious process of internalization of moral rules and norms to be changed into action. It thus helps create identification (*tongyi xing*) between self and society, and makes the 'natural person' (*ziran ren*) develop into a 'social person' (*shehui ren*) (Bakken, 2000).

This last point is crucial for the cultivation of collectivism which the Party leaders strongly advocate in the face of widespread individualism. It is argued that it is not enough to set boundaries, the process of taming modernity also depends on the bonds between individuals and their society, with individuals placing collective interests above their own (ibid.:104). Thus, self-sacrifice *(ziwo xisheng)* and selflessness *(dagong wusi)*, the highest principles in Chinese moral education both in history and during the Communist period (ibid.: 105), continue to be promoted in the reform era. Individuals are supposed to make necessary self-sacrifice for the overall social interests even when this means sacrificing their own lives. Therefore, self-sacrifice and even death for the common good are the main ingredients in the narratives about the exemplary model persons so widely used in the moral education texts. Moreover, these models all embody the virtue of 'hard work and plain living' (*jianku pusu*) and 'plain living and hard struggle' *(jianku fendou)*—two other old core concepts in the vocabulary of Chinese moral education, which the Party leaders believe can play an important role in fighting the 'limitlessness' of the 'self' today. As an extension of the discourse of collectivism, the Chinese government has taken pains to cultivate a strong identification with the existing Chinese nation-state, called 'patriotism' (*aiguo zhuyi*), among its citizens. In the 1990s, the government launched an extensive propaganda campaign to educate the people in patriotism. I shall elaborate on this and its implications for youth's identity construction in Chapter 8.

As we can already see, the construction of a 'new' theory of Chinese morality draws heavily upon traditional Chinese culture. Reflecting the aforementioned 'cultural particularism' embodied in the theory of 'socialism with Chinese characteristics', there has been a renewed interest in traditional culture among post-Mao social engineers. Government propaganda as well as the general intellectual debate overflows with analyses of the importance of traditional culture. For the Chinese leaders, this is not merely about a search for national identity in an increasingly globalizing world and hence resistance to globalization, but more importantly, it is about social control in the face of the danger of modernity. The inertial force of tradition has been directed against the dangers of modernity (ibid.). The construction of tradition in today's China is thus adopting a reflexive approach by taking one particular direction: searching for the binding and stabilizing elements in the tradition (ibid.). No doubt, the traditional Chinese culture, with its emphasis on harmony and balance, proves to be an especially rich resource for such a purpose. A wide range of values in this vein are referred to as 'traditional virtues of the Chinese nation' (*zhonghua minzu de chuantong meide*), which, according to the leaders, ought to be glorified, developed fully and carried forward. These include, *inter alia*, self-cultivation, self-control, self-sacrifice, collectivism, family harmony and happiness, loyalty, filial piety, modesty, humility, thrift, industry, education, discipline, and patience, just to name some.

It seems that the enjoinment of the Chinese modern subject to cultivate these moral qualities will well continue to accompany China on its journey in the new millennium for the goal of the 'harmonious society', which theoretically serves as an extension of the 'socialism with Chinese characteristic' thesis, particularly, the construction of 'socialist spiritual civilization'. One can clearly see that the ideal Chinese moral subject enjoined by the Party state stands at the core of President Hu Jintao's notion of '*ba rong ba chi*' (Eight Honors and Eight Shames), a summary of the so-called 'socialist notion of honor and shame', which often appears together with the 'harmonious society' slogan on banners all over the country:

- It is honorable to love the motherland; shameful to harm it.
- It is honorable to serve the people; shameful to betray them.
- It is honorable to follow science; shameful to be ignorant.
- It is honorable to be diligent; shameful to be indolent.
- It is honorable to be united and to help each other; shameful to make gains at others' expense.
- It is honorable to be honest and trustworthy; shameful to sacrifice ethics for profit.
- It is honorable to be disciplined and law-abiding; shameful to be chaotic and lawless.
- It is honorable to live plainly and work hard; shameful to wallow in luxuries and pleasures.

All this lends support to Bakken's (ibid.) argument that the Chinese society, in the past as well as today, constitutes an 'exemplary society', where the model of the ideal subjecthood is carefully constructed to guide human conduct. As is reflected in the ongoing discourse of the 'harmonious society', the exemplary society model may well continue into the future.

CONCLUSION

Dramatic transformation has taken place in Chinese society since the end of the 1970s, when China started its reform and opened up to the world with its pronounced blueprint of modernization. To set a general backdrop to the subsequent chapters, I have in the preceding pages offered an account of the major manifestations of this social drama and their impacts on social life. The chapter shows that social transformation under the auspice of 'socialism with Chinese characteristics' has entailed a series of apparent contradictions, contrasts and tensions, such as that between the old and new, individualistic values and collective norms, choice and risks, China and the West, wealth and poverty, material wealth and moral decadence and fragmentation and social cohesion. To understand the complexities inherent in such a social condition, it is important to recognize that continuity in cultural and political values and practices is concomitant to the change and that social transformation is not only reflected in economic and material development but also in the changing subjectivities and the accompanying subjectification regime in the reform era. I have shown that the reform and opening up has not only led to spectacular economic growth and improvement in people's living standard in general, but also a greater diversity of values and interpretations. There have been broadening and diversified contexts of action and expanded individual autonomy and choices, which has been highly celebrated by most Chinese, especially young people. Nevertheless, on the other hand, while it is embracing modernity, 'the late-socialist state, along with other actors, continues to condition the meaning of post-Mao autonomy through regulation of the domains in which and the ways that choices are made' (Hoffman, 2006a: 554). This conditioning has been made possible by the double subjectification regime in line with the ideology of China's 'dual modernity'.

Indeed, in the theory of 'socialism with Chinese characteristics' and through the discourse of *suzhi*, the two sets of seemingly contradictory subjectification technologies seem to have been successfully married as a coherent whole, thus having theoretically solved the inherent contradictions between them. Nevertheless, how the coexistence of conflicting values and enjoinments for the individual subject is being played out in the individual's lived experiences of China's modernization may not prove to be a smooth and easy process, as we shall see in this book.

I shall take up the themes that have emerged from this chapter further in the two following chapters, where I provide an analysis of the Chinese Internet and urban youth respectively.

2 The Internet with Chinese Characteristics

As indicated in the Introduction, the Internet does not constitute a globally uniform culture. Nor is it 'a placeless place', apart from the rest of social life. Therefore, the question is: What makes the Chinese Internet 'Chinese' and how. This chapter will cover the history of the Chinese Internet, rapid growth and remaining challenges, governmental control, competing discourses about the Internet in China and the emerging online culture in China. In so doing, I show how the Chinese Internet is being shaped by and also contributing to the shaping of the socio-cultural circumstances of today's Chinese society undergoing rapid and profound transformation.

RAPID GROWTH AND REMAINING CHALLENGES

As in many other countries, the academic and scientific sectors pioneered in developing and using the Internet in China. The first computer network was the China Academic Network (CANET) set up in 1987 to provide support for research in computer science by facilitating data sharing and e-mail exchange among distant researchers working on similar subjects. This was followed by other academic networks, such as the network of the Institute of High Energy Physics (IHEP) and the China Education and Research Network (CERNET). However, these earlier computer networks had no direct connection to the Internet. They were machine-to-machine links which later evolved into groups of machines interconnected via the X.25 protocol on public data networks. It was not until April 1994, when China's appeal for direct linking to the Internet was approved at the Sino-American Federation of Scientific and Technological Cooperation meeting in Washington, D.C., that China first got connected to the Internet. The first network directly connected to the Internet became active when the National Computing Facilities of China (NCFC), a collaborative network involving the Chinese Academy of Sciences, Beijing University and Tsinghua University, opened up a dedicated circuit to the Internet via Sprint Corporation, a foreign telecom firm, on April 20, 1994. In May 1995, China Public Computer Internet (Chinanet) began to construct its nationwide network and in January 1996, Chinanet opened its commercial Internet service to the general public.[1]

Thus, as we can see, in the early stages, only scientists, researchers and university faculty in the major metropolitan centers could access the Internet in China. It is estimated that there were only about 3,000 users by March 1995[2]. Since 1996, the Internet has shown exponential growth, which is not only eye-witnessed by native Chinese like myself, who have seen how digital devices enter Chinese people's everyday lives at an amazing pace. It has also been documented by the nationwide surveys by China Internet Network Information Center (CNNIC). As of October 1997, CNNIC's first Statistical Report on Internet Development in China disclosed the number of Internet users at 620,000. This figure climbed to 2.1 million by the end of 1998, 22.5 million in 2000, 111 million in 2005, and 298 million in 2008, according to CNNIC reports. In 2008 alone, the total represents a 56.2% year-on-year growth. The 25the CNNIC survey (2010), the latest available, shows that by the end of December 2009, China had 384 million Internet users, among whom 346 million (90.1%) were broadband users, and 16.8 million registered domain names. Affected by the launch of 3G (3rd generation) business, the number of mobile Internet users reached 233 million (60.8% of total Internet users), the total representing a 97.5% year-on-year increase. According to the same report, Internet users' online time continued to grow in 2009, with an increase of 2.1 hours per week per user. The average weekly online time grew from 16.6 in 2008 to 18.7 in 2009. The proportion of employed users accessing the Internet in their spare time is higher: 68.1% of users access Internet only in their spare time and 7.4% get online only at work.

Along with the rapid development, the penetration rate has also seen dramatic growth from only 1.7% in 2000 to 28.9% by the end of 2009 (CNNIC, 2010). Given China's huge population and compared with the penetration rates in some other countries, such as Korea (76.1%) and the U.S. (74. 7%) (CNNIC, 2009), the scale of Internet diffusion remains quite limited in China. However, the speed of development is most remarkable given that, by mid-1995, the country had only 3,000 users via the country's one international connection and that China's Internet boom appeared to be little affected by the worldwide IT industry slowdown at the turn of the century (Qiu, 2003).

In terms of places of access, CNNIC (2010) indicates that currently most users access the Internet via broadband at home (83.2%) and one-third (30.2%) at the office. Besides, as mentioned previously, Chinese Internet users are making great use of mobile devices (68.2%), which are ubiquitous in China, with cheap service and a great variety of phones available. It is worth mentioning that there are signs showing a decrease over the years in the number of people patronizing the Internet café, *wangba* in Chinese, as home and workplace Internet availability grows. Nevertheless, it seems that the *wangba* still constitutes a popular place for many urban youth, with

35.1% of users still accessing the Internet there. I shall explore the dynamics behind this phenomenon in Chapter 5.

The CNNIC reports show continuing improvement in equality in access along the structural dimensions of gender, age, education, income and socio-geography. For instance, the most remarkable change over the years is the narrowing-down of gender discrepancy. By the end of 1998, male users accounted for 86% of all Chinese netizens. This figure became 70% by the end of 2000 and 60% in January 2003. By the end of 2009, the proportion of male users to female users in China had been 54.2 : 45.8, relatively stable compared with late 2008. The change in the rural-urban ratio also appears impressive. Whereas by the end of 2000, only 0.76% of the Internet users came from rural areas that happened to hold 70% of China's population, by the end of 2009, the size of rural Internet users reached 106.81 million, accounting for 27.8% of total Internet users (CNNIC, 2010).

However, despite the rapid diffusion among other social groups over the past years, the digital divide in terms of access in China remains striking. As already indicated in the Introduction, currently the Chinese Internet is still mainly an urban youth phenomenon, which means most of the users are members of the only-child generation. Typical users tend to be young, urban, from affluent families, highly educated, unmarried and male. According to the 25th CNNIC report, which shows continuing, and sometimes exponential diffusion among other social groups, by the end of 2009, above one third of the users (31.8%) were 10–19 years old and almost another third (28.6%) were 20–29 years old. Only 17% were 40 years old or over. Although decreasing over the past years, the proportion of users with an education level of senior high school and above was still 64.5% in 2009. In the meantime, although income discrepancy had been gradually decreasing since 2000, the typical user was still more likely from an urban family, in most cases, China's emerging middle class family, which reflects the income discrepancies in China discussed in Chapter 1. It seems that the Internet in China will remain largely an urban phenomenon for some time given that the annual growth rate of rural users is lower than the overall Internet users growth rate nationwide (CNNIC, 2010). The 25th CNNIC report indicates that the scale of growth of Internet users in China is gradually slowing down, from 41.9% in 2008 to 28.9% in 2009. Although it is still too early to say for sure, this may mean that after universal adoption in the city, reaching a kind of saturation, there may be a general slowdown in growth given that the Internet may still be beyond most rural people's budgeting and relevance system.

Furthermore, it is worth noting that the 'digital divide' is a complex issue that should not be understood merely in terms of access. Disparities may exist among users in terms of equipment, autonomy of use, skill, social support and the purposes for which the technology is employed (DiMaggio and Hargittai, 2001). Particularly, the 'connectedness', which means whether or not and to what extent Internet connections will enhance people's chances of upward social mobility (Jung, Qiu and Kim, 2001), may vary among different social groups (and from person to person). Research shows that people who are already advantaged with respect to education, income, age (younger) and gender (male) are more likely to be connected to the Internet in ways that will preserve or increase their advantages (Jung, Qiu and Kim, 2001). This may well be the case in China. However, given that more detailed data of uses by various social groups are lacking in the CNNIC surveys and very little research exists that addresses this lacuna, further research is needed to gain a more nuanced and in-depth understanding of digital inequality in China.

GOVERNMENTAL CONTROL

An intriguing idiosyncrasy about the Chinese Internet is that the explosive growth and diffusion has been taking place within a context of stringent governmental control that many believed would handicap China's Internet industry from its inception (Lagerkvist, 2009; Qiu, 2003). An often cited study by the OpenNet Initiative (2005) found that China has the most sophisticated Internet control regime in the world. The control involves a combination of legal regulation, surveillance and punishment to promote self-censorship, as well as technical measures (Lum, 2006; Tan, 1999; Walton, 2001; Zittrain and Edelman, 2002).

Since 1995, when China began to construct its nationwide network, the Chinese government has issued extensive regulations regarding Internet usage, with the first one dating back to 1996. These regulations are directed at Internet service and content providers, Internet café operators and Internet users themselves (OpenNet Initiative, 2005). The year 2000 alone witnessed six major regulations regarding state secrets, online business operation, information and news services and Internet security (Qiu, 2003). For instance, the Measures for Managing Internet Information Services issued by the State Council in September 2000 makes it illegal for content providers to disseminate information which the government deems to be subversive and unhealthy. The Provisional Administrative Regulations on News Service by Internet Websites promulgated in November 2000 defines the content to be forbidden on Chinese news groups and news sites, and restrains the dissemination of information from foreign media. Other measures with more ambiguous legal status are also in force, such as the Good Conduct Charter imposed in March 2002 by the Chinese Internet Association on the major news portals and search engines, which reportedly had 130 signatories by 2003, including Yahoo China (Weber and Jia, 2007; Sautedé, 2003).

The types of information forbidden by the major regulations include:

1. That which is against the basic principles of the Constitution;

2. That which endangers national security, discloses state secrets, undermines state sovereignty, and injures national unity;
3. That which harms national dignity and interest;
4. That which provokes hatred and discrimination among nationalities and injures national solidarity;
5. That which undermines state religious policy and advocates cult and feudal superstitions;
6. That which disseminates rumors, disrupts social order, and injures social stability;
7. That which disseminates obscenities or pornography or promotes gambling, violence, murder, or terrorism;
8. That which defames or slanders others or impinges on the legal interests of others;
9. That which is otherwise prohibited by the law and administrative regulations (Zhou, 2006: 142).

It is prescribed that no group or individual may use the Internet to create, replicate, retrieve or transmit the kinds of information listed here. According to Lum (2006), at least 12 different government agencies nationwide, which form a monitoring and controlling pyramid, are involved in the implementation of Internet regulations.

Technically speaking, China censors the Internet through website blocking and key word filtering by means of its notorious 'Great Firewall', which blocks access to information deemed sensitive and undesirable by the government (Zittrain and Edelman, 2002), such as information related to Falun Gong (the Chinese-government-defined 'evil cult'), human rights, democracy, Taiwan independence and pornography. Thousands of websites (many porn related) are blocked outright, and destinations such as YouTube, Flickr and Wikipedia are heavily restricted (James, 2009). In addition, advanced Intranet and tracking technologies are also applied and these, together with the 'Great Firewall', are maintained by means of China's multi-billion dollar 'Golden Shield' project (Walton, 2001). Since 1998, dozens of intruders have been tracked down and imprisoned on charges such as disclosing state secrets, inciting subversion, Internet hacking, propagating Falun Gong and spreading rumors during the SARS epidemic (Qiu, 2003; Sautedé, 2003). And yet, as Randy James at Time.com rightly observes, generally speaking, the government's rules for what's permissible online are sweeping and, like much of its rhetoric, vague. News, for instance, should be 'healthy' and 'in the public interest'. Audio or video content must not damage 'China's culture or traditions'. And nothing must challenge the Communist Party and work against the building of a 'harmonious society' (James, 2009). Moreover, as James Fallows at *Atlantic Monthly* writes, the enforcement is often 'selective' too. This tactic is called 'killing the chicken to show the monkey (so as to teach the latter a lesson)' (*sha ji gei houer kan*), an old Chinese saying for ruling with intimidation. It is hard to know what, when or who will become the target of censorship. This may reflect the challenge faced by the Chinese government to juggle between having to keep the Internet open enough for the commercial benefits and to maintain control over it for social-political stability. However, more importantly, such strategies may even serve the purpose better, as they can lead to the most stifling restriction of all—self-censorship. In Fallows' words, 'The idea is that if you're never quite sure when, why and how hard the boom might be lowered on you, you start controlling yourself, rather than being limited strictly by what the government is able to control directly' (quoted in James, 2009).

Administrative measures are also utilized to ensure the implementation of the regulations. China has a branch of 'cyber police' (estimated at 30,000–40,000 members) working under the Ministry of Public Security (MPS). Established in 2000, the 'cyber police' is said to be becoming more and more sophisticated over time (Mooney, 2004)[3] and it operates at national, provincial and municipal levels. Besides investigating online crimes, such as spreading viruses and pornography or attempting financial fraud, the cyber police monitor websites and e-mail content and remove objectionable or subversive material (Lum, 2006). In addition, the government solicits help from citizens themselves to monitor Internet content (Lum, 2006; Lagerkvist, 2009; Qiu, 2003; Yang, 2009b). Those patrolling the Chinese Net not only include full-time cyber police but also the so-called 'state information security liaison personnel' (*guojia xinxi anquan lianluoyuan*), many of whom are college students who get subsidized for computer and Internet access by working part-time for law enforcement (Qiu, 2003). In June 2004, China Internet Illegal Information Reporting Center (CIIRC) was established jointly by the Ministry of Information Industry, the Ministry of Public Security and the State Council Information Office of the People's Republic of China.[4] It encourages citizens to report 'illegal' or 'harmful' information. As I shall show later on in this chapter, the government is also trying to capitalize on the Internet's capacity for propaganda by 'occupying' or 'colonizing' cyberspace.

It is noteworthy that, although it is important to recognize the constraints caused by the vigorous censorship, it would be overly simplistic to view the Chinese user as a mere helpless victim. As Yang (2009a) rightly notes, Internet censorship inadvertently forces users to higher levels of linguistic creativity. It is not uncommon that Chinese netizens devise new ways of bypassing the filters, such as by separating characters with hyphens and commas, using English acronyms or wholesale Romanization, or using Chinese characters with similar pronunciations to the forbidden words. To illustrate this, I cite the example of how Chinese users have 'twisted' the Chinese characters for 'harmonious society' (*héxié shèhuì*), Hu's ideal society mentioned in Chapter 1, to go around filtering and express their social and political satire. Since the Chinese leadership announced the goal of constructing a 'harmonious society', the government usually cites this as the reason for censorship. When the word for censorship was censored, Chinese netizens began to use the word 'harmonious' (*héxié*) as a substitute for censorship. And when the word 'harmonious' began to be censored, Chinese netizens began to use the word 'river crab' (*hé xiè*), as the word 'river crab' sounds similar to the word 'harmonious' in Chinese Mandarin. Sometimes aquatic product (*shuichan*) is used in place of 'river crab'. It is also used as a verb, for example, instead of saying something has been censored, one may say 'it has been harmonized' (*bei héxiéle*) or 'it has been river crabbed' (*bei hé xiè le*).[5] For example, not long ago, in our chatting, one of my Chinese students told me that she found the Chinese Internet filtering mechanism increasingly ridiculous because it does not distinguish between the really 'bad' content from the 'harmless'. A few of her postings, apparently not harmful at all, 'were "harmonized"' (*bei héxiéle*), she complained. However, notwithstanding such instances as these, generally speaking, China's Internet control seems to enjoy wide support among Chinese, both users and non-users, as I shall discuss further later on in this chapter.

Chinese government's unyielding control of the Internet has to do with the fact that from the very beginning, the technology has been viewed as a driving force for China's national economic development, rather than citizens' civil participation and democratization, as I shall show in the next section.

BETWEEN EXTOLLING AND DEMONIZING: THE CHINESE NARRATIVE OF THE INTERNET

With a strong belief in science and technology as a most powerful means of national development, the post-Mao leaders share the understanding that information technology is the locomotive of modernization of the country (Dai, 2003). Since the 1990s, the government has been fully engaged in building an information society through a digital revolution. Informatization of the national economy and society has been incorporated as a strategic priority in the national tenth 'five-year plan' (2001–2005) (*China Daily*, October 8, 2003). The top leaders have reiterated their unambiguous view on the importance of information technologies for China on numerous occasions. For instance, former President Jiang Zemin famously remarked that 'None of the four modernizations (modernizations of agriculture, industry, education and the military) would be possible without informatization.'[6]

The post-Mao leaders' prioritization of digital technology is based on their belief that it will enable China to achieve its goal of 'technological leapfrogging' (*jishu kuayue*) (Dai, 2003; Damm, 2007). In 1983, soon after China started its reform and opening up, former premier Zhao Ziyang claimed, 'The new technological revolution or information revolution . . . may help China skip over some of the stages which have been experienced by other developing countries' (Hamrin, 1990: 213). Similarly, Premier Zhu Rongji stated in 2000 that 'leapfrogging in productivity development may be achieved—by melding informatization and industrialization, the two processes reinforce each other and progress simultaneously' (Dai, 2003: 8). The strong belief in information and communication technologies among the leaders has been much influenced by the discourse produced outside China. For instance, the discussion of China's 'technological leapfrogging' often draws on the work by the American futurologist writer Alvin Toffler. His book, *The Third Wave,* which was published in 1980 and translated into Chinese in the early 1980s, became a bestseller in the country, and so did his other books later on. In the book, Toffler describes the transition of developed countries from Industrial Age society, which he calls the 'Second Wave', to the Information Age, that is, the 'Third Wave' society. Chinese leaders used the term 'technological leapfrogging' to refer to the idea of omitting, or leapfrogging the industrial phase with the help of information and communication technologies to catch up with Western developed societies in the Information Age. This is reinforced by the host of academics and international organizations that have urged developing countries to take advantage of the global communications revolution to jump-start economic development, emphasizing that the economic future will belong to the technologically adept (Dai, 2003: 8). The goal of 'technological leapfrogging' serves as the major motivation for China's state sponsoring of the Internet, with a strong investment in the development of the infrastructure and technological innovation, which helps make the—still highly controlled—Chinese Internet the fastest growing in the world (Dai, 2003; Damm, 2007; Qiu, 2003).

China's strong motivation to initiate a digital revolution also has to do with forces inside the country, particularly, the country's own past (Dai, 2003; Qiu, 2003; Tsui, 2005). Accompanying Chinese leader's ambition of 'technological leapfrogging' is a strong sense of humiliation for the atrocities inflicted upon the nation since the Opium War (Qiu, 2003). Such an emotional complex underlies China's whole discourse on technology and modernization. Thus, an important component of the Chinese national narrative about the information revolution, and China's modernization in general, is the collective remembering (Wertsch, 2002) that China used to be the most technologically advanced nation in the world, yet it fell from the throne in its recent history and suffered tremendously partly due to the rulers' rejection of advanced technology. The Chinese discourse about modernization, thus, revolves around a wish to 'restore' power to the Middle Kingdom by re-achieving its technological supremacy. Reviving the Middle Kingdom through advanced technology has been a constant goal for Chinese leaders since Sun Yat-sen, who gave up his presidency of the Republic of China to become the Minister of Railways out of a belief in the power of technology. By the same token, just as the Maoist regime attempted to emulate Soviet industrial statism through frenetic social movements such as the 'Great Leap Forward', the post-Mao government has been enthusiastically sponsoring the country in its informatization revolution since the 1990s. The post-Mao leaders have insisted that since China reacted too late to the Industrial Revolution, resulting in its backwardness on the world's stage, it cannot afford to miss the Information Revolution, which provides China with another, but rare opportunity, to catch up with its East Asian neighbors as well as the developed world (Taubman, 1998). Unlike the miserable failure of Mao's Great Leap Forward, however, the digitalization campaign centered on the Internet has been largely successful by a range of indicators, such as growth, sustainability and the capacity of the Chinese state to boost the information industry via investment, purchase, assistance in R&D (research and development) and the acquisition of foreign capital (Qiu, 2003).

In the discourse about the Internet and China's technological leapfrogging, there is an inherent extolment of the new medium's power, whereby the Internet is construed as hopeful, benign and empowering. Nevertheless, the Chinese government's view of the Internet is far from that simple. Reflective of the government's ambivalent attitude toward modernity as both signifying hope and danger for China, as discussed in Chapter 1, the Internet, which is an ultimate symbol of modernity, is viewed as a double-edged sword. It serves as an embodiment and a means of modernization key to China's development. Meanwhile, however, it provides access to information and social spaces the government deems not conducive to the social-political order (Golub and Lingley, 2008; Yang, 2003a). The ambivalence about the new medium underlies the government's tremendous efforts in controlling the Internet while relentlessly promoting the technology's growth. Moreover, governmental concern about the new medium's 'bad' influence on social order reinforces and is reinforced by a widely shared Internet anxiety among the general public, especially urban parents. Chinese urban parents not only share the larger concern about the Internet's 'bad' influences on the younger generation's morality, hence Chinese society's moral order. More importantly, they tend to see the Internet as highly counterproductive to their children's academic achievement, which is assessed mainly through textbook-based exams in the exam-oriented education system, despite their wish to keep their children up-to-date in the use of new digital technologies in line with the discourse of the 'information society'. I will discuss the widely shared Internet (café) anxiety in China in further detail in Chapters 5 and 6.

The Chinese state's concern about the erosion of social norms by new, foreign-imported technology can be traced back to the eighteenth century when, defeated by the West in the Opium Wars, China started to rely on imported technology for modernization of the country (Tsui, 2005). Tsui (2005) makes the point that the notion of '*tiyong*'—which is the shorthand form of 'Chinese learning for substance, Western learning for practical use' (*zhongxue wei ti, xixue wei yong),* and which emerged in the mid nineteenth century as an ideological solution to the question of how to conserve Chinese values while applying Western technology—still guides the Chinese approach to new technology, in particular, the Internet. That is, reflecting its general dual approach to modernity, the Chinese government wishes to both harness the economic benefits of the new medium and to fight against its potential harm for the established social-political order. This is why as with other forms of mass media in China (Zhong, 2002), the Chinese state refuses to totally loosen its grip on the Internet. It shows that just as the Party leaders believe that they can conquer the 'juggernaut' of modernity (Bakken, 2000), they are determined to tame the 'monster' of the Internet. And for the Party leaders, both rely on a successful disciplining of the body and mind of Chinese people.

Besides the legal, technical and administrative control strategies, the government is actively exploiting the propaganda capacities of the new medium itself to mitigate the real and potential influences from foreign cultures (Weber and Jia, 2007; Zhou, 2006). That is, as Zhou (2006) observes, along with suppression and regulation, the government also uses subtler and softer strategies to achieve its goal. For instance, the Chinese state has consistently attempted to employ the media to reinforce the uniqueness of Chinese values and practices (Cao and Downing, 2008; Weber and Jia, 2007). Through a National Online Game Publishing Project, the state has been encouraging the development of domestic online games bonded to traditional culture and socialist model figures. According to SAPP (State Administration of Press and Publication, 2004), the state was sponsoring 100 domestic online games between 2004 and 2008, the aim of which was to 'strengthen . . . moral and ethical education' and 'cultivate healthy consumption' among young people. This was guided by the propaganda principle of 'educating youth through entertaining them' (Zhong, 2002). One such example of the Chinese government attempting to promote political-moral education in a manner that might appeal to technology-savvy youth is the marketing of computer games that substitute Communist model figures such as Lei Feng and other historical heroic figures for heavily armed superheroes and reward players for doing good deeds instead of gunning down evildoers. However, the response reportedly has been mixed, with some saying 'cool', but not in the positive sense (*Global Technology Forum*, November 9, 2005[7]).

Other proactive ideas and measures are adopted as well to guide and influence the production of public opinion in online forums, such as bulletin board systems (BBSs). A new mechanism of 'Internet commentators' (*wangluo pinglun yuan*) was introduced in 2004 (Yang, 2009b). According to Yang (2009b), these include both volunteers and paid staff, whose mission is to covertly guide the direction of the debate in accordance with the principles laid down by the propaganda departments of the party, rather than to promote critical debate. The story goes that provincial police units in several provinces have been trying to achieve this goal by employing the so-called '*wu mao dang*' ('the 50-cent Party') for posting counter-comments to 'negative' opinions. These people, usually young and Internet savvy, have acquired this derogatory name because they are given 50 Chinese fen in payment for each 'positive' counter-posting they make in an online forum (Bandurski, 2008; Lagerkvist, 2009; Yang, 2009b). It is estimated that these paid commentators amount to as many as 280,000 people nationwide (Bandurski, 2008; Lagerkvist, 2009).

Such efforts as mentioned here support a value system that continually evokes the uniqueness of the Chinese race, Confucianism and socialist values. In doing so, domestic media groups reinforce the notion of an 'imagined Chinese community', linked economically to, but separated politically and culturally from, the global world (Weber, 2003). This brings us back to the basic principles of China's 'dual modernity' under 'socialism with Chinese characteristics'. The Chinese government's attempts to reign in online behavior and speech for the sake of social-political order are in keeping with its double subjectification regime tailored for its new subjects.

FROM 'TECHNOLOGY OF FREEDOM' TO 'TECHNOLOGY OF CONTROL': THE WESTERN NARRATIVE OF THE CHINESE INTERNET

For a long time, there has been a Western narrative about the Chinese Internet that differs fundamentally from how the Internet is discursively constructed inside China (Damm, 2007). The introduction of the Internet into China, especially the Internet boom at the turn of the millennium, has aroused intense interest in the West in (watching) how the new medium is going to change the political system by bringing about democracy to the country. The general anticipation has been that the Internet, as an 'uncontrollable' form of technology, will give voices to the Chinese citizens, set them free from the authoritarian regime and dismantle the state apparatus (Damm, 2007; Kluver and Yang, 2005; Tsui, 2008). Then it follows that by adopting the Internet, the Chinese government is digging a grave for itself, or 'unwittingly letting in a Trojan Horse' into China's current political system (Tsui, 2008). Political elite in the U.S. led in constructing this narrative about the Chinese Internet. Former U.S. president Bill Clinton, for instance, famously stated that attempting to control the Internet in China is like trying to nail Jell-O to a wall.

Underlying such discussions about the Internet in China is the notion of the Internet as a 'technology of freedom' (Damm, 2008), which has been a widely shared perception, rarely challenged until recently. This kind of technological determinism has its root in how the medium was conceptualized in its earlier years. As Bucher (2004) observes, two well-known metaphors have informed the notion of the Internet as a 'technology of freedom'. One is the 'information superhighway', coined by the former Vice

President of the United States, Al Gore, in his widely cited speech on January 11, 1994, at the University of Los Angeles. This metaphor emphasizes connection, public access, speed and free flow of information. The other metaphor is 'cyberspace', a term coined by the science fiction author William Gibson in his sci-fi novel *Neuromancer* (1984). With this metaphor, the principles of linearity, speed and access control, which stand behind the highway metaphor, are replaced by the principles of openness, non-linearity and free communication and interaction (Bucher, 2004). These features of 'cyberspace' are best captured by John Philip Barlow (1996) in the 'Declaration of the Independence of Cyberspace'. Barlow explicitly claimed, 'Cyberspace does not lie within your borders. Do not think that you can build it, as though it were a public construction project. . . . It is an act of nature and it grows itself through our collective actions.' For him, cyberspace is 'global social space', 'the new home of Mind' and it 'consists of transactions, relationships, and thought itself'. And he anticipates 'a world that all may enter without privilege or prejudice accorded by race, economic power, military force, or station of birth.'[8] In short, the metaphor of 'cyberspace' implies the end of geography, history and politics, as Mosco (2004) has argued.

Another assumption underlying the Western narrative about the Chinese Internet concerns the Chinese people. As Tsui (2008) argues, such a narrative is in line with the Cold War mentality that assumes an oppressed people starved for uncensored information and awaiting to be set free by the 'truth'. According to this assumption, Chinese people are naturally opposed to the government, which reflects what Wang (2001: 1) sees as a Western propensity to view Chinese politics as 'a constant division and tension between 'the authorities', which are preconceived as hostile to the West, and 'the people', who must be friendly to the United States and its allies.' It follows then the encounter between the Internet, the 'technology of freedom', and the Chinese people, the oppressed, is like a hungry man seeing bread or a fish on the land coming back into the water. Such assumptions neglect the fact that most Chinese people support the government's effort in controlling the Internet, as I have mentioned previously and as I shall show in more detail in the next section, and that they are quite critical about the Western media and their interest in the Internet in China only when it concerns matters of censorship (Damm, 2007; Rosen, 2009; Tsui, 2008).

As increasing evidence points to China's successful control of the Internet even as it continues to open itself up to the world, the dominating Western discourse about the Chinese Internet has largely lost its utopian tone and taken up a dystopian view. The discourse about 'liberation' is now largely replaced by that of 'control', which is in fact another version of the Internet as 'technology of freedom' discourse (Damm, 2007). The main concern shared by politicians, commentators, the media and academics now is that state censorship and regulation may prevent the Internet from realizing its 'liberating' potential. There is a tendency in the West to interpret every single action taken by the Chinese authorities as a 'crackdown' (Damm, 2007). Metaphors and terms used to describe China's Internet control include 'the Great Firewall', 'the Panopticon' and 'the Orwellian Internet'. These metaphors, all of which signify a seamless control, fail to capture a more comprehensive and realistic picture of the Internet in China. As Yang (2009a: 111) argues:

> Although the Internet is also under control, its networked features provide numerous nodes of resistance against control. Furthermore, business interest in a prosperous Internet economy serves as a buffer against control because an overly controlled Internet will keep users away and harm commercial prosperity.

To sum up, as Damm (2007) rightly argues, the problem with the Western narrative lies not in the facts that are reported. Rather, it is the narrow focus that is problematic. Focusing only on censorship and crackdowns disregards the much wider societal implications of the Internet, as pointed out by an increasing number of scholars (e.g., Damm, 2007; Kluver and Yang, 2005; Tsui, 2005, 2008; Yang, 2009a; Zhou, 2006). It reflects a technological determinism which tends to 'omit human intention, practice, and social institutions that mediate technology and any sort of political outcome' (Yang, 2009a: 109). In so doing, it also tends to ignore the mundane Internet use, hence the emerging Internet culture in China.

THE MUNDANE ONLINE LIFE AND THE EMERGING INTERNET CULTURE IN CHINA

In this section, I shall discuss the emerging Internet culture in China by offering a brief account of how Chinese users tend to use *their* Internet. Here I use the term 'Internet culture' interchangeably with 'online culture' or 'cyber culture' and in its broadest sense to signify the features that have been emerging from the use by Chinese netizens of the medium for various purposes. These features, thus, together with the aforementioned Chinese narrative of the Internet, make the Chinese Internet distinctive from the Internet in other socio-cultural contexts.

Implicit in the Western narrative about the Chinese Internet is the assumption that Chinese users are either busy inventing ways to circumvent the government's censorship measures in pursuit of the 'forbidden truth' or are left at a loss in front of the Great Firewall. Empirical findings convincingly show that this assumption largely falls short of the realities of the everyday online lives of Chinese netizens. Large-scale surveys over the past decade such as the semi-annual reports by CNNIC and the studies conducted by Guo Liang and his colleagues at the Chinese

Academy of Social Sciences (CASS) consistently place entertainment at the top of the list concerning Chinese users' purposes for going online. For this reason, Guo (2005) argues that the Internet in China has become the 'entertainment superhighway' rather than the 'information superhighway' as it is called in the West. For example, the latest report by CNNIC (2010) shows a picture not substantially different from those presented by the previous CNNIC reports. That is, a high proportion of Chinese netizens use the Internet for entertainment-oriented information: 83.5% use it for online music, 68.9% for online games and 62.6% for online video. Also popular is online communication, with 70.9% of netizens using the Net for instant messaging, 56.8% for e-mail, 57.7% for blogging and 30.5% for BBSs. The report further shows that 80.1% of the users surveyed reported using the Internet for news and 73.3% of them use search engines. By contrast, use rate for business transaction remains low, showing that Chinese netizens trail behind their Western counterparts in this respect (although business transaction application sustained the fastest growth among all the use categories in the latter half of 2009) (CNNIC, 2010). This general trend is confirmed by the large-scale studies by some international research teams, such as the 'Young Digital Mavens' project by the IAC and JWT (2007), two advertising companies, and the Boston Consulting Group's study of China's digital generation (Meyer, Michael and Nettesheim, 2008).

An even more entertainment-oriented use pattern emerges if one takes into account the fact that the distinction between the categories in the earlier-mentioned studies, especially that between entertainment, communication and information, can be misleading. What is blurred by these distinctions and the figures is that online communication and online entertainment are often interwoven in the Chinese case. For instance, much of the instant messaging takes place between gamers during games, as I observed during my China fieldtrips and is indicated by studies on gaming in China (e.g., Cao and Downing, 2008; Lo, 2009; Wu et al., 2007). Besides, reading blogs, and even writing blogs, is also often used as a form of recreation (Wang and Hong, 2010). Search engine use can also be much about entertainment-related information. According to CNNIC (2009), as of June 2009, China was home to 235 million search engine users, that is, 69.4% of all users, up by 59.49 million from June 2008. However, music was the most commonly searched form of non-text content, with 39.5% of users having searched for music. Movies, videos and games followed, with 30.3%, 13.7%, and 16.5% of users respectively. Likewise, search for text-based content can also be about information or news about game websites and online game magazines, as indicated by *China Online Games Market Report* (2007–2008),[9] and celebrities, as I learned from my interaction with Chinese young users.

Chinese netizens' preoccupation with online entertainment and social interaction constitutes one of the main reasons behind the rapidly growing game industry in China. CNNIC (2010) reports that in 2009, the number of online game users reached 265 million, an increase of 41.5% compared to 2008. Niko Partners (2010), the leading market intelligence firm on China's video game industry, in its *Annual Review and Five-Year Forecast on China's Video Game Industry*, indicates that while the global economic downturn hurt video game publishers in much of the world, China's online game industry was not negatively impacted in 2009 and gamers continued to embrace online games as the best inexpensive source of social entertainment available. The most exciting feature of the Chinese online game market is the explosion of MMORPGs (massively multiplayer online role-playing games), which got their foothold in China around 2000 and soon dominated the market, dwarfing other forms of video games (Cao and Downing, 2008). A major reason for their rapid development in China is that such games combine entertainment with social interaction as well as offer opportunities for identity experimentation (Cao and Downing, 2008; Lo, 2009; Niko Partners, 2010; Wu et al., 2007). All this may prove to be especially attractive to urban Chinese youngsters, who, with no siblings, tend to be thirsty for peer-to-peer interactions and who are typically eager for individual self-expression, as I shall discuss next and in Chapter 3. A good example is the popularity of in-game marriage, called web marriage (*wanghun*) or virtual marriage (*xuni hunyin*) in some of the Chinese MMORPGs (see *Global Times*, July 16, 2009; Lo, 2009; McLaren, 2007; Wu et al., 2007). By 2007, a reportedly one million people had experienced a web marriage (McLaren, 2007). '*Tianya* marriage hall', 'No. 9 City' and 'Online Family' (*wang shang renjia*) are some of the most popular web marriage sites. The number of such sites is still increasing, such as 'i-part', 'lovefree.chinesegamer.net', 'game.china.com/audition' and 'city-6.com', indicating that China's online marriage boom is still going on (Lo, 2009). These commercially driven sites for entertainment, pleasure and networking also constitute domains for self-expression, fantasy and identity experimentation in ways that both challenge and perpetuate traditional heterosexual norms (McLaren, 2007; Lo, 2009; Wu et al., 2007).

Another striking feature of Chinese online culture is that there is a predominant focus on personal life, lifestyles and consumerism, as is centrally reflected in Chinese users' online social production. Social production here refers to the process whereby users contribute to the Internet economy by producing non-proprietary content online, such as blogs and BBS postings (Benkler, 2006). Like online gaming, these serve as spaces for experimenting in self-expression and identity construction. Take the lively Chinese blogosphere as an example. Blogging has been enjoying vast popularity among Chinese users since 2002 (Wang and Hong, 2010). Many Chinese users got to know about blogging through the blogs of the young journalist Mu Zimei, who shot to fame in 2003 after starting a blog filled with descriptions of her sexual interactions with numerous men, and Sister Fu Rong's blogs, where she offers narcissistic narratives about her 'sexy' body and general 'personal charm', and shows her signature pose the 'S-figure', in which she bends her knees and thrusts out her chest provocatively. These two leading bloggers became a focus point for both the online cultural debates and the Party state's regulatory measures against 'unhealthy content' on the Internet in 2004 and 2005, which made the two names even more widely known (Farrer, 2005). China's number of bloggers has experienced rapid growth since 2002, from 0.28 million, of whom 0.23 were 'active bloggers', to over 16 million, of whom 4.3 million were active bloggers in 2005 (CNNIC defines active bloggers as those who update their blogs at least once a month). The year 2005 was also recognized as China's Year of the Blog due to both its popularity and its cultural impact on China's society (Yu, 2007a). By the end of 2009, the number of blog users had hit 221 million, accounting for 57.7% of Chinese netizens. This represents an increase of 59.40 million users and 3.4% of utilization rate compared to the end of 2008. The proportion of active blog users continued to grow, with the number of users who updated their blogs within six months increasing to 145 million, an increase of 37.9% (CNNIC, 2010).

China's Blogger Surveys (CNNIC, 2007, 2008) show that the most popular blog content is personal diary (84.9% in 2006 and 88% in 2007), which far outnumbers the second most popular category, information about literature, TV/movies/the arts and interest/hobby. The 2007 survey further reveals that the majority (83.5%) of Chinese bloggers write just to document personal feelings and experiences and 60.2% express personal views. Bloggers also employ the medium for humor, with content about humor increasing from 9.8% in 2006 to 16% in 2007. Large numbers of blog audiences (52.4% in 2006 and 43% in 2007) read blogs just to pass time. Moreover, blogs of friends and relatives are the most frequently visited, with a total of 68% of the audience reporting they frequently visit blogs of friends and relatives in 2007. Moreover, celebrities' blogs are highly attractive to China's blog audience. In 2006 and 2007, respectively, 24.5% and 19% of audiences responded that they frequently visited blogs of actors, actresses or celebrities. In terms of audiences' blog preferences, 57.2% of them in 2006 and 42% in 2007 liked to read about personal feelings and experiences, and 47.2% in 2006 and 43% in 2007 frequently read blogs related to emotions. Romantic relationships, the new category of the 2007 survey also gained 20% popularity among blog readers. In short, although there have been cases where bloggers have employed the blogosphere as a means to demand accountability of power holders on behalf of non-Internet users, political bloggers are relatively few compared to bloggers writing about entertainment, sports and celebrities (Lagerkvist, 2009).

A similar pattern seems to characterize the social production at BBSs. BBS remains one of the most dynamic social spaces in China today while electronic bulletin boards have almost become a thing of the past in the U.S. (Yang, 2009a). In China, BBSs are not only provided by the big portals, but also by numerous public, private and semiprivate websites as well

(Damm, 2007). BBSs enjoy high popularity among Chinese users, as thanks to many-to-many communication, such forums provide an ideal place for discussion and building online communities. CNNIC (2010) shows that by the end of 2009, one third (30.5%) of Chinese netizens reported having used this form of online forum, which is similar to 2008. Exactly because of the many-to-many communication and the popularity of BBSs in China, scholars (e.g. Giese, 2006; K. Liu, 2004; Tai, 2006; Yang, 2003a) have been interested in exploring to what extent or whether or not such forums constitute public spaces for Chinese netizens in the Habermasian sense, whereby participants engage in equal and free discussions about various questions concerning politics and identity. Whereas this remains a question for ongoing and future research, preliminary evidence—for example, the analysis made by Damm (2007) of the contents of Chinese BBSs—indicates that the BBS sites in China are for the most part concerned with consumerism, lifestyle and private life rather than with politics. One notable exception is the Strong Nation Forum (*qiangguo luntan*), hosted by the *People's Daily,* one of China's major official newspapers, which has served as a site where participants discuss various social-political issues. However, while serving as a means for people to let off steam about various social-political ills in China, the discussion is also much in line with the Party state's ideology. At times, the most dominating discourse in the forum is nationalism (K. Liu, 2004; Zhou, 2006; Qiu, 2003; Xu, 2001).

In short, there has been a sustained fascination among Chinese netizens with online recreation and communication over the past decade and a half (Yang, 2009a). Moreover, Chinese users' online social production is characterized with a strong interest in consumerism, lifestyles and identity politics. In general, what is usually lumped as online 'entertainment/recreation' (and communication) serves a wide range of needs for the user such as fun seeking, sociality, self-expression and identity experiment, but rarely is meant for subverting the government. Via such 'apolitical' online practices—and despite the effective censorship—the Chinese users seem to be (comfortably) living a dynamic and rich digital life. Indeed, it has been observed that Chinese users live a considerably more intensive digital life than their Western counterparts. For example, the aforementioned 'Young Digital Mavens', a survey of 16- to 25-year-old U.S. and Chinese youth by IAC and JWT (2007) concludes that Chinese youth lead their U.S. counterparts in digital self-expression, displaying a greater drive for creativity and experimentation with identity (see also Lagerkvist, 2009). The study shows that almost five times as many Chinese as Americans surveyed said they had a parallel life online (61% vs. 13%); 86% of the Chinese vs. 42% of the U.S. respondents agreed that 'I live some of my life online'. More than twice as many Chinese respondents (69%) agreed that 'I have experimented with how I present myself online' (vs. 28% of Americans). More than half the Chinese sample (51%) said they had adopted a completely different persona in some of their online interactions, compared with only 17% of Americans. As many as 82% of young Chinese agreed, 'Interactivity helps create intimacy, even at a distance,' compared with just 36% of young Americans. Almost two thirds (63%) of Chinese respondents agreed that 'It's perfectly possible to have real relationships purely online with no face-to-face contact' vs. 21% of Americans. This seems to indicate that the Chinese youth live a more 'expressive' online life compared with their Western counterparts. My own qualitative study comparing high school students in China and Norway confirms this finding by showing that the Chinese youth tended to see the Internet as a separate world from the 'real' world (as I shall show further in Chapter 4), in contrast with their Norwegian counterparts, who saw the Internet as a part of 'this world' (Liu, 2010).

The features of Internet use in China may have to do with the general new media culture in China, which is mainly consumption oriented (Damm, 2007; Qiu, 2003), despite the official rhetoric that frames the information technology as a means to promote productivity (Zhao, 2007). As Qiu (2003) observes, from the beginning, going online has been advertised as fashionable entertainment. As a highly attractive segment in the Chinese consumer market, the only-child generation becomes the major target of online recreation. Moreover, Chinese users' interest in more 'serious' forms of information seeking may also have been dampened by China's sophisticated Internet censorship. More importantly, however, the general characteristics of China's Internet culture are closely related to who the users are (Damm, 2007) and it has to do with the socio-cultural context of present-day China (as discussed in Chapter 1). As indicated previously, although a variety of different groups of Chinese Internet users can be identified, they share some common characteristics: urban, young, only-children and highly educated. Their strong interest in 'playing' and 'socializing' online reflects the general tendency among children and young people to engage in 'playful' online activities and socializing, which is common also in the West (Sandvig, 2006). Meanwhile, however, as I shall demonstrate in some of the later chapters, Chinese young users' preoccupation with online recreation and communication has to do with the only-child's social-biographical situation and relevance structure. With the well-known pressure on the only-child, as I will discuss in detail in Chapter 3, urban children and youth seem eager to carve out a social space via online recreation to satisfy their subcultural needs and cope with the pressure in daily life. This tendency is reinforced by a yearning for social connection in times of individualization and social fragmentation concomitant to the profound social transformation in China.

Their general 'apolitical' orientation toward the Internet is also closely related to the fact that despite in-group disparities due to the restratification of Chinese society in the reform era, these technological 'haves', mainly urban youth, largely belong to the new urban middle class. As Damm (2007) has rightly argued, they are a beneficiary of the economic, and to some extent, political policies of the late 1990s and the new millennium. They constitute an important part of a consumerist postmodern society, with a strong interest in personalized and individual lifestyles. As I shall discuss further in Chapter 3, much less politicized than urban youth in the 1980s and as the beneficiary of the reform, this group has a highly pragmatic approach toward the government and the Chinese Communist Party (CCP) (Rosen, 2004, 2009; Damm, 2007; Yan, 2006). They generally support the government's modernization project and the control of 'sensitive political issues' and 'unhealthy' online content and practices (Damm, 2007; Lagerkvist, 2009; Tsui, 2008). The findings from a series of surveys about Internet use in China from 2000 to 2007, which were funded by the New York-based Markle Foundation and directed by an internationally respected research team at the Chinese Academy of Social Sciences, indicate that the majority of the respondents, both users and non-users, support governmental control of the Internet. According to the fourth such survey (Guo, 2007), over 80% of respondents (both users and non-users), said they thought the Internet should be controlled, and almost 85% said that the government should be responsible for doing it. Concerning what needs to be controlled, 87% of Internet users would control pornography; 86% violent content; 83% spam or junk mail; 66% advertisements; 64% slander against individuals. Fewer respondents targeted the very popular but less malicious entertainment and recreation opportunities. Half of respondents said online games should be controlled, and more than one in four (27%) said online chatting should be controlled. The findings also show that since 2005, the percentage of users who said that online content about 'politics' should be controlled or managed jumped from 8% to 41%, by far the biggest increase of any items tested.

Of course, the general lack of interest in politics that characterizes the Chinese Internet culture should not be taken to mean that the Net is a mere playland or shopping mall. Guobin Yang (2009b), in his book *The Power of the Internet in China*, convincingly shows that online activism constitutes an important venue through which the social issues attracting most widely shared concern get voiced and contended. The main issues handled by such activism cover not only popular nationalism, but also right defense, corruption and power abuse, environment, cultural contention, muckraking and online charity (Yang, 2009b: 55). Often spontaneous in nature, online protests happen as a countermovement against social injustice and inequality (Yang, 2009b; Lagerkvist, 2009). Examples range from efforts to free hundreds of children and mentally handicapped men who had been kidnapped and forced to work as slaves in brick kilns to protests that halted the construction of a dangerous chemical plant in the port city of Xiamen. There was also an outpouring of support from the online public for the artist and blogger Ai Weiwei in his endeavor to draw society's attention to the issue of bad school construction and the children killed in those schools during the Sichuan earthquake.[10] In 2009 alone, a number of neologisms emerged from Chinese netizens' active participation in online protests and discussion concerning a series of incidents of corruption, social injustice and inequality, such as 'hide-and-seek' (*duo maomao*) '70 yard' (*qishi ma*), 'pressure differential' (*yali cha*), 'impromptu quality' (*linshi xing*), 'the passive tense era' (*bei shidai*) and 'snail-dwelling tribe' (*wo ju zu*)[11]. In this way, netizens are demanding accountability, transparency and public participation from their government (Yang, 2009b; Lagerkvist, 2009). As such, despite its normally 'apolitical' appearance, the Chinese Internet has also occasionally served as an effective platform from where demands of accountability from power holders are made (Lagerkvist, 2009). However, it is noteworthy that reflecting the post-1989 social-political conditions, online activism rarely demands radical political change. Rather, the struggles are about social justice, citizenship rights, cultural values and personal and collective identity (Yang, 2009b). Moreover, compared with the mainstream 'apolitical' uses, such uses count as only a small fraction of the Chinese online life. Nevertheless, it still constitutes a strong and emerging social force that is contributing to social change, a force that the Chinese state cannot afford to ignore (Yang, 2009b).

That the Chinese Internet is not a mere playground or shopping mall is also reflected in the fact that vast numbers of online communities exist in Chinese cyberspace. According to Yang (2009b), online communities are perhaps the most important new associational form to have emerged since the 1990s. Numbering in the millions, they exist in newsgroups, mailing lists, chat rooms, BBS forums and blogs (Yang, 2009b). Online community building, such as that in Chinese cyberspace, constitutes an identity movement, expressed as yearnings and struggles for social recognition, personal dignity and a sense of community, and ultimately, for a better world (Yang, 2009b). I will take these points up again in Chapter 7, where I examine the collective identity work going on in an online community run by and for contemporary Chinese youth and in the concluding chapter. Moreover, the Internet has also been used for other 'more serious' purposes, most notably, during times of crisis. For instance, it has served as an effective platform for information and mobilization during SARS (Yu, 2007b) and the Wenchuan earthquake in 2008, as anyone who followed the event on the Internet would have found out.

Furthermore, it is important to note that not all online entertainment is 'apolitical' either, just as Yang (2009b) emphasizes. Online 'play', such as spoofs—for instance, the famous political satire 'A Bloody Case Caused by a Steamed Bun' (see Yang, 2009a and Yu, 2007a for analyses of this)—constitutes less overt forms of challenging conventional political and cultural power, just like blogs such as those by Mu Zimei and Sister Furong.

CONCLUSION

Situating the Chinese Internet within its socio-cultural context of 'dual modernity', this chapter shows how the Chinese Internet has come to acquire its Chinese characteristics. The discussion highlights a few striking features of the Chinese Internet which often take the form of apparent contradictions and paradoxes. First, the rapid and sustained growth in online population has made China the country with the greatest number of netizens in the world. However, just as China's spectacular economic growth has been accompanied by striking social disparities, its information revolution has benefited some while marginalizing others. On the whole, reflecting the social stratification discussed in Chapter 1, the urban middle class, especially youth, constitute the technology 'haves', whereas their rural counterparts the 'have-nots'. Enjoying access to the Internet is one of the most important elements of urban youth's everyday lifeworld, which is both indicative of their privileged social position and serves as a major source of societal and parental concern, themes that I shall deal with more systematically in the following chapters. Secondly, rapid growth has been taking place within the context of strict control. Whereas this may appear puzzling at first sight, a closer scrutiny shows that this is not unlike the miraculous economic growth within the context of the authoritarian one-Party rule. Controlling the Internet has been part and parcel of the social control deemed necessary by the state for China's success in its modernization project, or in realizing 'socialism with Chinese characteristics'. The analysis of the Chinese national narrative of the Internet shows that the Chinese government's attitude toward the Internet, which is an ultimate symbol of modernity, exactly reflects its general dual approach toward China's modernization. As the Internet, like modernity in general, stands for both hope and danger for China, ambivalence characterizes the Chinese attitude toward the new technology. Thus, the state leads an anti-Internet discourse, joined by the public, particularly parents, at the same time as it exalts the Internet as an empowering tool enabling China to achieve its goal of 'leapfrogging'. The Internet anxiety thus constitutes part of the larger concern about the moral crisis in the reform era as well as changing notions of success and hence, the parental fear for the Internet's distraction of young people from schoolwork and other activities necessary for success (Golub and Lingley, 2008). As such, the Chinese understanding of, and approach to the Internet, differs fundamentally from the Western narrative about the Chinese Internet, rendering the latter largely irrelevant in the Chinese context.

A third feature related to the Chinese Internet is that despite China's sophisticated censorship regime, a dynamic and lively online culture is emerging from the mundane uses by Chinese netizens. Chinese users manage to display—often via what is usually lumped in the Chinese discussion about young people and the Internet as 'online recreation/entertainment'—a drive for creativity and experimentation with self-expression and identity, which seems strikingly strong compared with, for instance, their American and Norwegian counterparts (IAC and JWT, 2007; Liu, 2010; Lagerkvist, 2009). The analysis of the emerging Internet culture in China also indicates that although predominantly focused on entertainment and communication (often mixed), Chinese netizens can also, although in relatively rare cases, turn the Internet into a platform for more serious purposes, such as political activism and social mobilization during acute events and for social justice. This shows that despite stringent governmental control and the predominant tendency toward entertainment/communication, the Chinese Internet is not merely about entertainment and control—'two misleading images medial stories and survey reports tend to perpetuate' (Yang, 2009b: 1). Nevertheless, despite such cases of more 'serious' uses of the Internet, the entertainment-communication orientation seems to predominate in Chinese cyberspace, reflecting a youth culture with a focus on consumerism, fun seeking and the 'cool', as I shall show in the next chapter.

The idiosyncrasies about the Chinese Internet can best be understood by locating the Internet in its local socio-cultural context and users' social-biographical situations. In other words, the major features observed about the Chinese Internet are the joint effect of the ideology of 'dual modernity', with its concomitant double subjectification regime, and users' active appropriation of the new medium in ways that reflect their offline actualities. As such, the analysis in this chapter lends support to Yang's (2009a) argument that the Internet as a new cultural form, while making possible creative expressions and representations, necessarily reflects new social conditions of the reform era.

Whereas the social transformation depicted in Chapter 1 serves as the more general backdrop to the empirical chapters as well as this chapter, the Chinese Internet with its Chinese characteristics, as described in this chapter, serves as the immediate micro cultural environment in which young people's interaction with the Internet is being played out. The major themes that have emerged from the analysis in this chapter will thus be taken further in the empirical chapters, where I explore young people's perceptions and experiences of the Chinese Internet as well as their online self-representation.

This chapter has briefly touched upon the social-biographical situations of urban youth, the main group of China's netizens. However, to further contextualize my informants' interaction with the Internet, I shall devote the next chapter to an account of who today's Chinese urban youth are.

3 Paradoxes as Lived Experiences of Modernization

Urban Youth with Chinese Characteristics

People may develop distinctive generational characteristics due to rapid social change as well as through traumatic historical events (Edmunds and Turner, 2002; Mannheim, 1952/1923). Present-day Chinese youth (using the broader definition of youth mentioned in the Introduction) represent China's first generation whose lives have from the very beginning been accompanied by the country's reform and opening up with no lived experience of Maoist socialism. Thus, their identity construction is being played out in a dramatically different socio-cultural context compared with the previous generations of Chinese youth. In addition, as already mentioned, due to the one-child policy instituted in 1979, nearly all of urban children and youth are only-children. A range of expressions have been used to highlight the impacts of the dramatic social transformation upon Chinese urban youth, such as 'the newer, new humanities' (*xin xin renlei*), 'China's millennials', 'China's me-generation', 'China's generation Y', 'China's Internet generation' (alternatively, 'China's e-generation' or 'wired generation'), the 'post-80s generation', 'little emperors/empresses' and 'the new radicals'.

Obviously, such expressions used to describe a whole generation of youth across a whole society run the risk of neglecting in-group differences, or 'generational units' in Mannheim's (1952/1923) terms. Chinese urban youth today are far from unified in terms of their living conditions, belief systems and behaviors, which reflects the increasing pluralisation of Chinese society after thirty years of reform (Rosen, 2009). For example, clear differences are said to exist between the 'post-80ers' (*balinghou*, those born 1980–1989) and the 'post-90ers' (*jiulinghou,* those born 1990–1999) (although both are sometimes put under the general category of the post-80ers'). And many youth themselves seem to identify with the differentiation between the two (sub)groups, as is evidenced by the ongoing 'online cross-firing' between the two groups (*China Daily*, July 8, 2008; W. Zhang, 2008) and as I shall show further in Chapter 7. Nevertheless, it seems still valid to draw some general patterns of the collective character of present-day Chinese urban youth, which will be the task of this chapter. I will touch upon a few interwoven themes such as their living conditions, life chances, struggles and frustration as well as well-being, and their value orientations.

THE PRIVILEGED GENERATION

Chinese urban children and youth today are known to enjoy material prosperity and well-being that would have been unimaginable for previous generations. They have never experienced the hardships of civil war, the turmoil of the Great Leap Forward or the Cultural Revolution and the famines and the shortages that their parents and grandparents lived through. Moreover, the material comfort they enjoy is often comparable to that in which their counterparts in developed societies find themselves (Fong, 2004b; St-Maurice and Wu, 2006). As indicated in Chapter 1, the sustained double-digit growth in gross national product (GNP) thanks to the radical economic reforms launched at the end of the 1970s has led to much greater purchasing power of urban families. Due to the particular pattern of social stratification in reform-era China, generally speaking, urban families constitute the main body of the newly emerging middle class in China, characterised by family material prosperity and consumerist lifestyle.

Moreover, their only-child status tends to enhance urban youth's material well-being. It has been well documented that the one-and-only child occupies a central position in the family, where parents tend to concentrate all of their resources on the one child (Fong, 2004a; Liu, 2002a, 2006a; Rich and Tsui, 2002; Woronov, 2002). In addition, each of these 'little emperors and empresses', as they are called, has two pairs of doting grandparents (maternal and paternal grandparents). It is not uncommon in such families that one finds the unique phenomenon of 'one-family two-tiered consumption'—with parents spending less on themselves and using lower-tier brands in order to save for their child, who purchases premium brands (Chen, 2005; Yueh, 2006). It is estimated that urban parents spend an average of 50–70% of the total household's income on their only-children for their needs or desires, which range from education to designer brands (Lu, 2005; Sun, 2003). A study by the Shanghai Academy of Social Sciences shows that urban children's education expenses alone entail 46% of their families' total expenses at the lower secondary stage, 51% at the upper secondary school and 52% at the university level (see *China Daily*, April 1, 2005). In addition, most urban children have their own discretionary savings. According to a study by the China Youth and Children Research Center, between New Years' gifts and allowances, urban teenagers have an average of over 200 yuan a month of their own money (*People's Daily*, October 5, 2006). All this makes urban children and youth a major target of marketing in terms of educational, fashion and recreational consumption (Gong, Li and Li, 2004).

Nourished by the widespread consumerism in China's reform era and the wealth they enjoy, today's urban youth are known to be brand conscious and fashion-savvy. Many of them grew up consuming brand goods such as Nike and Adidas and frequenting McDonald's and Kentucky Fried Chicken (Jing, 2000)—Western fast foods, the consumption of which has become a symbol of material modernity and new cultural expressions in China (Yan, 2005). A McKinsey study shows that 65% of Chinese teenagers feel that it is very important to keep up with the latest fashions and that 85% indicate that they prefer name brand clothes and mobile phones (St-Maurice and Wu, 2006; see also Lim, 2008). This is reinforced by the fact that Chinese youth are strongly affected by peer pressure and believe it is important to buy certain products to compete with their classmates (Chee, 2000; Fong, 2004a). A touch of hedonism encouraged by the rampant consumerism is often associated with this generation (K. Liu, 2004; Weber, 2002), in sharp contrast with the previous generation who lived a 'hard and plain' (*jianku pusu*) life under Maoism.

This generation is also media savvy, well informed about cutting-edge global trends and enamored of cool communication gadgets, constantly thinking of upgrades (Lim, 2008; Wang, 2008). They live in a world of mobile phones and computers, Hollywood movies, NBA games, English pop songs and Japanese comic books. The pop, rock and roll, blues and jazz on their MP3 players and iPods seem to be much more appealing to them than the revolutionary anthems and they quickly develop a taste for hip-hop (Yan, 2006). As the pivotal group of users of new ICTs in China, they are wired to the World Wide Web (albeit often through the circuits of governmental control)—a gateway to places, thoughts, ideas and news the likes of which their parents couldn't even imagine when they were young. This means that Chinese youth today have unprecedented access to diverse channels of information and knowledge compared with previ ous generations.

Education is another major area in which today's urban youth draw the admiration of their parents, who were largely deprived of the opportunity to take formal education beyond lower secondary school. The vast majority of Chinese urban youth today have had 14 or more years of education (that is, upper secondary school or its equivalent) and learned at least some of a foreign language (English in most cases) (Yan, 2006). As I shall show next, as only-children, their lives are closely associated with (aspirations for) higher education.

Chinese urban children and youth today are also known to enjoy a more democratic relationship with their parents (Liu, 2006a; Sun, 2003) compared with earlier generations. The social transformation described in Chapter 1 has greatly affected parent-child relations in the urban Chinese family.In sharp contrast with the past, a child-centered culture has emerged as a result of the combined effect of the rarity of children due to population control, decreased desire for fertility concomitant to modernization, improved family economy and greater exposure to modern ideas of child rearing (Liu, 2006a). Parents increasingly tend to view their children as equals or friends and encourage a more interactive relationship (Rosenberg and Jing, 1996; Sun, 2003). This change has sometimes taken an extreme form, with parents of only-children abandoning the authoritarian parenting style with which they themselves grew up and adopting a highly permissive and even pampering attitude toward their only-children (Liu, 2006a). In line with this child-centered culture typical of the only-child family, research further suggests that urban children are quite influential in the selection of many family products and services, including food, vacations, clothing and restaurants (McNeal and Wu, 1995; McNeal and Ji, 1999; McNeal and Yeh, 2003; Ying, 2003). Although this may be exaggerated in the media, as suggested by Flurry and Veeck (2009), my judgement is that they do exert considerable influence on family consumption choice.

In short, present-day Chinese urban youth find themselves in the center of material comfort, rampant consumerism and broadened educational opportunities and accesses to information. Compared with the previous generations and with their rural counterparts, especially those in the remote and minority areas, they enjoy advantaged social-economic-cultural conditions that promise for better life chances (Jacob, 2006). Moreover, as China's youth with 'global identities', they tend to have the same high expectations, consumption patterns and educational achievements as their counterparts in wealthier societies (Fong, 2004b). In this sense, it can be argued that they are the privileged social group in China. This constitutes the material and cultural foundation for the widely observed individualism among urban youth, a topic that I will take up in a later section of this chapter.

THE GENERATION UNDER GREAT PRESSURE

The preceding section conveys a picture of Chinese urban youth enjoying prosperity and well-being and being complacent. Nevertheless, this does not conceal the pressure and frustration which constitutes an equally salient theme in the life stories about this generation (Fong, 2004a, 2004b; Liu, 2006a, 2008a, 2008b; Yan, 2006). With the shift from a planned economy to market economy, urbanites no longer enjoy the so-called 'iron rice bowl', which refers to the socialist welfare package, including jobs, life-long employment, housing, medical care, retirement pensions, and even arrangements for their children's jobs. Thus, despite the expanded individual autonomy in the reform era, gone is the sense of security enjoyed by urbanites a few decades ago. They are supposed to create their own 'rice bowls' like all others in society (Bray, 2005). Furthermore, the rice bowl may prove hard to create given the severe employment situation (as I shall show later) and the scarcity of educational and professional opportunities in China (Giles, Park and Cai, 2006; Ghose, 2005; Song, 2003). Moreover, even if one has succeeded in creating a rice bowl, it may prove hard to fill due to the changed notion of what counts as a 'good' life. As I show in Chapter 1 and later on in this chapter, Western middle-class lifestyle has become a norm among urbanites, especially young people who view it as *the* good life. Such a notion puts great pressure on young people (supported by their parents) to strive to accumulate capital in various forms, be it economic, social or cultural (Liu, 2006a). The pressure is reinforced by the impending filial duty expected of the only-child in a society where the state has failed to provide a sound security system for the elderly (and the young), both within the only-child family and beyond (Ikels, 2006; Liu, 2008b; Rich and Tsui, 2002; Stockman, 2000; Zhan and Montgomery, 2003). One of the emerging issues related to the only-child family concerns the so-called '1-2-4-plus' formula, which means that during their working years a couple may have to support one child and four parents, and in some cases, up to eight grandparents as well, without the help of any siblings (Festini and de Martino, 2004). Traditional filial values are still greatly encouraged by the state in present-day China and supported by legislation (Ikels, 2006; Palmer, 1995). Research shows that most Chinese young people, including young adults who are members of China's only-child generation, internalize this value (Croll, 2006; Deutsch, 2006; Liu, 2008b; Whyte, 1994).

Within a Confucian tradition that stresses the importance of education, the insecurity, changed notions of the good life and impending filial duty, together with the improved financial capacity of the only-child parents, have led to the widely observed high parental achievement expectations in the urban community (Fong, 2004a; Liu, 2002a, 2006a, 2008a; Rich and Tsui, 2002). It has been a well-known phenomenon that from the very beginning of life, the only-child seems to be tied up with education and particularly, there seems to be a strong relationship between the only-child family and higher education (Croll, 1995; Fong, 2004a; Liu, 2002, 2006a, 2008a; Rich and Tsui, 2002). Research (Liu, 2002; 2006b) shows that only-child parents of both boys and girls suffer from what I call 'the university complex', whereby they are preoccupied with getting the child into a university, preferably a prestigious one. Such an eagerness for the child's university education is constantly accompanied with hope, worry and anxiety. Starting from the very early years of the child, the daily lives of the whole family, especially that of the child, are heavily framed by how to reach that goal (Fong, 2004a; Liu, 2002; 2006a). However, even when one has got a university education, it is not deemed to be enough. In the earlier studies I have conducted (Liu, 2002, 2006a), the only-children, both young adults and younger ones, and their parents share an understanding that a bachelor's degree is no longer enough today to secure 'the good life'. Therefore, those who already had a seat at university did not dare to stand still; most of them deemed it necessary to take postgraduate studies (Liu, 2008a). This extended 'university complex' has been reinforced by the inflation of educational diplomas in the age of mass higher education, whereby it is only the credentials from prestigious institutions that seem to count. This means that, reminiscent of what Brown (2003) calls 'the opportunity trap', in realizing the broadened range of 'choices' or 'opportunities' in the neoliberal market economy, few can afford to opt out of the competition for a livelihood and meanwhile the expectation of such an investment seems hard to cash out. Those who stand at advantage, in this case, urban youth, 'are having to run faster, for longer, just to stand still' (Brown, 2003: 142).

But it seems that the academic goals of the only-child family are not easy to achieve given the selectiveness of the exam-based, stratified education system and the great patience and strong will it requires to endure all the long years of hard and monotonous school work. Chinese education has undergone a fundamental transformation since the reforms began in the late 1970s. With the reform, the previous re-distributive social-political aims of education became subsumed under national economic development goals. Accordingly, Maoist egalitarianism was rejected and the system of 'key-point' schools and universities reintroduced. Higher education, for instance, has been classified into a hierarchical system, in which key-point institutions are identified and strengthened, all institutions ranked, and key majors favored—all of it encouraging stratification within the system (Yin and White, 1994).

It is the exam system that works as a fateful mechanism to stratify young people into different social categories and echelons. The two most decisive examinations, *zhongkao* (Upper Secondary School Entrance Exam) and *gaokao* (National College Entrance Exam), taking place at the end of lower secondary school and upper secondary school respectively, to a great extent determine not only who is placed where in the hierarchical education system but also who is struck off the academic track. Due to the importance attached to the two public exams, young people's secondary school years are mainly spent preparing for them, which leaves little free time for students to pursue their own interests. Many schools practice a military type of discipline and control of students in order to direct students' total attention to school work so that they can score as well as possible in the exams, which has implications for both the school's status as an institution and teachers' career promotion and material benefits. It is common that secondary school students have a lot of home assignment to do in addition to having classes from early in the morning till late in the evening, six or seven days a week (Liu, 2006a). Moreover, extracurricular activities such as taking after-school classes and hiring private teachers are common practices among the urban community (Liu, 2006a; Wang, 2007). Likewise, the great majority of Chinese parents have strict control over the television watching habits of their offspring (Chan and McNeal, 2003) and they control their youngsters' use of the Internet out of concern about the Net's distracting effect on their children's academic performance, as I will show in further detail in Chapters 4–6. This shows that the more democratic relationship between parents and children in the only-child family hardly applies to this domain of their lives, but mainly to consumption choice.

The Chinese process of schooling is demanding in still another sense. Like in the past, the current exam system is still mainly a paper-and-pen test of the students' grasp of the textbook contents. Thus, although the Confucian *Four Books* and *Five Classics* have been long replaced with a range of modern subjects such as mathematics, physics, chemistry, Chinese, English and politics, the exam still relies heavily on book knowledge and rote learning. Methods such as memorization, imitation and repetition, which have dominated Chinese education for centuries, still represent the order of the day (Bakken, 2000). Partly because of this, the Internet, despite its potential for innovative teaching and learning, has rarely been adopted as an important means of education so far, neither in most Chinese schools nor in higher education institutions (Qiang, 2007). As I shall show later on in this book, the Internet is often seen as the antithesis of learning. In order to score high on the exams, students are buried in 'a sea of exercises' of a fixed type for several years until they have taken the two major exams (Gao, 1998; Liu, 2006a).

All this sheds light on why for the vast majority of secondary school students, school is boring and exhausting, no matter whether one is a 'good' student or otherwise (Gao, 1998; Liu, 2006a; Yan, 2006). A 2007 survey of 1,400 urban families by the Youth and Children Research Center indicates that most Chinese youngsters feel exhausted by the burden of homework and extracurricular activities (Wang, 2007). In a 2005 nationwide survey (cited in Yan, 2006), 67% of the teenagers said they always worried about exams, and 41% feared they might not be able to enter a prestigious university. Of the primary school students and middle school students surveyed, 66% and 77%, respectively, reported not having enough sleep. In response to the question 'What would you like to do if you have free time?', 53% of the teenagers chose 'to have a long and good sleep.'[1] Although they only spent about five hours in bed per night, some of them still asked the doctor to help them find a way to sleep even less so they could spend more time studying and preparing for exams. In the earlier-mentioned study on the only-child generation (Liu, 2006a), my informants indicated that their greatest wish at school was to have a vacation, where they would not do anything but sleep and play. According to statistics from the China Psychology Association (cited in Yan, 2006), over 10% of Chinese under the age of 17 do not have sound psychological health because of heavy schoolwork, employment pressure and a monotonous life. Most tragically, ranked as the fifth leading cause of death among the Chinese population, suicide is the major cause of death among China's youth.

Nevertheless, most urban youth today manage to complete lower secondary school and many do make it to the end of upper secondary school (Connelly and Zheng, 2003). This is in line with the high parental educational expectations and the pragmatism widely found among Chinese urban youth, as I shall discuss later on in this chapter. Yet, despite the widely shared dream about the university among the urban community and the long years of frustration, it is still a relatively small number of the age cohort who manage to get into higher education institutions, especially those that are good enough to count in the inflation of academic credentials related to the massification and commercialization of higher education. The goal of enrolling 15% of the age cohort was achieved in 2002 and by 2006 the gross enrollment rate of the 18–22 age cohort reached 22% according to official statistics (MoE, 2007). But selectivity still persists. In most cases, vocational schools, many of which are almost purely commercial and hardly reliable for necessary training leading to employment, become the final resorts for the vast numbers of the 'educational discards' upon graduation from the lower and upper secondary school. This not only reflects urban parents' wish that their only-children can have as much education and training as possible and their ability to pay for this. Moreover, they also need to find an institution where the youngster can mature further rather than going 'loose' in the street, given the difficulty to find a job, especially one the only-child cares to do (Liu, 2006a).

Pressure and frustration are not limited to teenagers, but also apply to older young people, be they university students and graduates, or non-student young adults. Surveys among Chinese college students show that about 40–45% worry about their prospects of finding good jobs, and a small number begin to worry about car loans and mortgage payments even before they graduate. Most intriguingly, some of the older '80ers, who are still in their twenties, have shown concerns about being outdated and replaced by the '90ers (see Yan, 2006). Worse still, a 2004 survey of 1,010 college students in the city of Wuhan, central China, reveals that 11% of the students had thought of committing suicide; a 2005 survey of 200 students in a Beijing college shows that 30% of the students had the idea (see note 1 in this chapter). This has much to do with the severe employment situation for young people. With the dramatic expansion of Chinese higher education, employment for graduates has become a serious social problem in the country, which has been a hot topic covered by the media during the season of graduation. China's Ministry of Education statistics indicate that China had some 7.1 million graduates seeking jobs nationwide in 2009, with 6.1 million fresh graduates, whereas in 2002, the total number comprised only 1.45 million (*China Daily*, January 22, 2010). It was estimated that with the economic crisis exacerbating the problem, close to 2 million graduates were not able to find work in 2009—many of whom were postgraduates, even doctoral graduates (Zhou and Lin, 2009). According to an earlier report by *People's Daily* (March 7, 2005), in Beijing graduates with doctoral degrees lined up for high school teaching positions. Ph.D. holders intending to settle down to a high school teaching career are becoming increasingly common in many high schools in Beijing. One such school received 105 resumes from candidates with doctoral degrees, and more than 3,000 resumes from master degree holders. Whereas those who have failed to enroll in higher education may now comfort themselves with the thought that university diplomas may after all not matter that much given the predicament faced by college graduates, the situation for themselves may not be more sanguine, either. Most of such youths can only engage in temporary jobs in low-payment sectors. Finding a job or establishing one's own business that is sustainable is equally hard for non-university youths. In the aforementioned study I conducted on urban youth (Liu, 2006a), the non-university participants, with similar life aspirations to their university counterparts, were constantly worrying about losing their jobs or getting squeezed out of a certain business they had just entered. They were forced to constantly think about updating their qualifications and creating ways to make money while finding both hard. Some simply 'chose' to rely on their parents for a living. Currently, the term 'double-loss youth' (*shuang-shi qing nian*) has become a popular expression among Chinese youth. It refers to youth who lose both their job and lover almost at the same time under the global financial crisis. Many young people get dumped by lovers after losing their job because they refuse to maintain an unemployed boyfriend or girlfriend.[2]

Chinese urban youth's unemployment or postponed employment contributes to a phenomenon in Chinese cities similar to what is referred to as 'emerging adulthood' (Arnett, 2000) in some developed countries. One should be cautious about treating only-children as a homogeneous group, given the social stratification in the reform era both within the same city and across geographical areas, as indicated in Chapter 1. Nevertheless, it seems not an overstatement to say that 'emerging adulthood' is a widely prevalent phenomenon among the urban community (Nelson and Chen, 2007). The fact that an increasing number of urban youth in their twenties still have to rely on their parents for basic living because they could not find satisfactory jobs and afford their own independent homes has earned them the title '*kenlaozu*', meaning literally the 'parents-chewing-tribe', or 'those who bite their parents (for a living)' (Yan, 2006). This phenomenon is currently being reinforced by the dramatic rise in prices of home apartments in Chinese cities, especially big ones such as Beijing, but also more generally. This has been a hot topic in the Chinese media.

Being able to have an 'emerging adulthood', however, does not necessarily mean one's life is happy. On the contrary, the 'emerging adulthood' is a major source of frustration, pressure and boredom for many Chinese urban youth. As downright pragmatists, as I shall show later on in this chapter, and caught between the temptation of the middle-class lifestyle and the lack of employment and professional development opportunities, they are supposed to be rational, instrumentalist subjects who can put off enjoying life in order to live *the* good life in the future. Those who do not manage to practice such self-control may find themselves constantly condemned by others and feeling uneasy themselves.

But their frustration and the strong sense of pressure originate from still another source. Present-day Chinese urban youth tend to cherish an admiration for wealthy Western societies based upon media representations of life abroad and global interaction (Kwong, 1994; Fong, 2004b; K. Liu, 2004). They seem to be constantly comparing China with those societies, which leads to a sharpened dissatisfaction with the perceived inadequacies of Chinese life (Fong, 2004b). For example, Fong (2004b) shows in a study based upon 27 months of participant observation in several secondary schools and a survey of 2,273 secondary school students in Dalian City, one of the most developed cities in China, that the urban youth she studied were constantly frustrated by China's 'backwardness' compared with the affluence of wealthier societies. Well-educated, raised on wealthier societies' images and brand names and viewing themselves as the vanguard of modernization in their families and society, they identified with an imagined global community of wealthier societies, rather than with the 'backwardness' of China, where, they believed, it was their misfortune to be born. Fong further observed that this dissatisfaction with their own country does not necessarily mean that they lacked a sense of filial nationalism for their motherland. This kind of frustration may be especially keenly felt by young people in the less developed areas given the actual and perceived limitations of life in such places.

Struggling for upward mobility either within national boundaries or, preferably, within the imagined community of global elite (or both) (Fong, 2004b), which is strongly urged by their parents typically with high expectations, Chinese urban youth may feel the pressure under which they are living to be too heavy. This constitutes a striking theme in the narratives of the young people I interviewed in the earlier study (Liu, 2006a) and those in this study. Moreover, as the generation raised in an era of diverse, often contradictory cultural influences, who were socialized to believe that they are embodiments of high quality (Fong, 2007), Chinese urban youth today may often find it hard to communicate with the older generations, who hold more conservative values and have had less education (Liu, 2006a, 2010; Marr and Rosen, 1998). In addition, although most of them internalize high parental expectations and cherish a due respect for their parents' authority within the family in accord with the traditional filial piety, some do find it hard to come to terms with the grand plans of their parents, whose expectations can at times prove stifling (Liu, 2006a, 2010).

To sum up, the widely shared sense of great pressure among Chinese urban youth constitutes the other side of their apparent well-being and broadened range of choice and possibilities described earlier on in this chapter, which reflects the general socio-cultural conditions of today's China. Together, these different aspects of their lives form a striking paradox characteristic of Chinese urban youth. As I shall show in the following pages, paradoxicality equally applies to their value orientations, which is closely related to the material and cultural realities of their everyday lifeworlds discussed so far in this chapter.

THE ME-GENERATION

What I have shown in the last section indicates that present-day Chinese urban youth are indeed China's 'new humanities', living their lives in dramatically different circumstances than their forbearers. However, what distinguishes them even further from the previous generations, lending further support to their 'newness', is their strong individualistic tendency. For this reason, they have been referred to as China's 'me-generation' (e.g., Elegant, 2007). This is not surprising given the child-centered family culture and the general socio-cultural context of post-Mao China, where several interrelated forces have contributed to an individualization process, as discussed in Chapter 1.

The kind of individualism that China's me-generation embraces with alacrity is succinctly captured in these comments from Yunxiang Yan (2006: 255) on the commercial 'I'm lovin' it', which, according to Yan (2006), was designed by several Chinese youth in 2003 for the fast food giant McDonald's and spurred a surprising boost in sales:

> In its Chinese form, the slogan *wo jiu xihuan* (I'm lovin' it) conveys a much stronger, imposing, and disaffected tone and thus sends out more effectively a me-centered message: I am simply loving it; who cares what you think! Such a message, with the image of a hugely successful Taiwanese pop singer sitting on the hood of a red Jeep, acutely resonates with the life aspirations of the current generation of Chinese youth who are determined to achieve instant individual gratification in terms of a pleasurable and comfortable material life.

As is clear from this portrayal, Chinese urban youth's individualism has a double emphasis on individual freedom and material achievement. As such, it incorporates both an eager pursuit of individualistic self-expression and an entrepreneurial spirit with a heavy dose of materialism. These two forms of individualism come close to what the authors of the book *Habits of the Heart* (Bellah, et al., 2008) refer to as 'expressive individualism' and 'economic or utilitarian individualism'. Fluck (1998: 59–60) offers a useful distinction of the two different manifestations of individualism as follows:

> In the traditional form of economic individualism, . . . self-esteem is derived primarily from economic success and social recognition. In order to obtain these, the individual has to go through an often long and

> painful act of deferred gratification and self-denial. Analogous to the act of saving, the goal is to accumulate a stock of capital, in both economic and social terms, which will eventually yield its profits in the form of increased social approval and a rise in the social hierarchy. . . . In contrast, the culture of expressive individualism is not primarily concerned with a rise of the individual to social respectability or its (tragic or melodramatic) failure, but with the search for self-realization. Its major issues are no longer economic success or the promise of social recognition, but the assertion of cultural difference, that is, the ability of the individual to assert his or her own uniqueness and otherness against the powers of cultural convention and encroaching disciplinary regimes. If development and growth are key terms of economic individualism, difference is the key term of expressive individualism.

Bellah, et al. (2008) use these two terms to represent the two major stages of the individualization process in American society, by which they mean that the one form has dominated over the other in different periods of history with utilitarian individualism being gradually replaced by expressive individualism (Fluck, 1998). However, in the Chinese case, it seems that these two forms of individualism are simultaneously and equally forcefully displayed by the cultural identity being forged by the younger generation, which is closely related to the rapid and dramatic social transformation in the reform era. I shall further discuss the two forms of individualism among Chinese urban youth in the two following subsections.

THE 'KU' GENERATION: CHINESE URBAN YOUTH AND EXPRESSIVE INDIVIDUALISM

Chinese youth's preoccupation with expressive individuality is centrally reflected in the popularity of the slang term 'ku',[3] the Chinese version and phonetic translation of the English slang term of approval 'cool'. As the most representative among the popular slangs used by present-day urban youth, 'ku' endures through its association with the new individualism youth identify with (Moore, 2005). The interest in the 'ku' is so prevalent among Chinese youth that this generation has been referred to as China's 'generation ku' (Moore, 2005).

A quite precise explanation in Chinese of the term 'ku' is found at the 'encyclopedia' (*baike*) at China's biggest search engine Baidu.com.[4] I draw on this source for the purpose of this section. According to this site, the English slang 'cool' was introduced into Taiwan in the 1970s, where it was translated as 'ku', meaning 'being unrestrained' or 'elegant and unconventional with a degree of indifference'. In the 1990s, the expression was taken up in mainland China via the mass media, rapidly replacing the classical expression '*xiaosa*' with similar meaning and became the most popular praising words among youths. The term has been constantly enriched in its meaning over the years and now it generally means 'good', instead of merely meaning 'elegant and unconventional with a degree of indifference'. One will say 'ku' when wishing to express admiration for a person or the person's behavior, or simply for a thing. During an interview with one of my informants in a previous project on China's only-child generation,[5] she indicated that she did not want to have so-and-so as her boyfriend, because, 'he is not ku'. Believing that she herself is a 'ku' person, she insisted that she must have a 'ku' guy as her boyfriend. '*Ku Bile*' (literally being 'ku' to death or extremely 'ku') is used when one extremely admires something or somebody. Nevertheless, although it generally means 'good', the 'ku' in youth's eyes differs substantially from the traditional sense of 'good', as the earlier-mentioned article in the Baidu 'encyclopedia' aptly points out. When young people utter 'ku' to express admiration for a certain person, this person must be one who shows individuality in his or her manner of dressing, words, actions, demeanor and temperament. By no means is it the 'good' in the traditional sense about a person, who acts according to convention and shows qualities such as being 'simple and honest' and 'cordial'. Rather, 'individuality' and 'independence' are the essence of 'ku'.

An account of the 'newer new humanities' at zhidao.baidu.com talks about 'ku' as follows, which can serve as a further illustration of what it means to be 'ku' for Chinese youth:

> No other term is like 'ku': the newer new humanities are never tired of using it. Ku represents the fashionable, the avant garde, the rebellious, the other, indifference, soberness, the stimulating and the fun. It also represents a kind of wisdom, wisdom in constructing one's own independent life and taste. . . . Perhaps the current main stream society may feel that the newer new humanities' thoughts and behavior are irrelevant and far-fetched. However, when the newer new humanities become the main stream of society in the future, the ku culture, ku life, ku work, ku sports and ku art that they advocate will become the ku fashion of the main stream society.[6]

It is noteworthy that, as Moore (2005: 358) states, 'the ku of China's millennials is not a carbon copy of Western styles'; there are different ways to be ku in contemporary China. Nevertheless, it clearly represents an individualism that is much influenced by Western pop culture and the values it references, with an emphasis upon modernity and individuality. We can get a glimpse of the type of expressive individualism embodied in the term 'ku' through a look at the newer new humanities' view on 'fun'. According to the earlier-mentioned account of the 'newer new humanities', for them, 'it is none of your business' (*bu guan nishi*) is an ardently embraced motto and 'the most important is to have fun' (*zui yaojin shi haowaner*), which is followed by the following elaboration:

> The things that the newer new humanities find fun are usually things even the mention of which suggests stimulation and adventure: climbing, bungee jumping, free-style snowboarding, camping alone, high-speed surfing, and so on. Golf? No. How decrepit and old-fashioned! Bowling? No. Isn't it too 'white-collaredish'! Kara OK? No. How boring! The newer new humanities do not like these. . . . Under the flag of individuality, they like to do things that ordinary people do not like to do. These things are full of challenges and risks. Life is so short. Therefore, perhaps, one really needs to be a bit crazy in order to make it more fun. . . . Of course, having fun is not limited to sports. Such a spirit of pursuing fun permeates newer new humanities' all realms of life such as study, work, lifestyle, entertainment, pastime and so on. Only fun contents can attract their interest and make them engage in them fanatically and whole-heartedly.

This seems reminiscent of their counterparts in the West, who are individualistic fun-seekers and who want to go beyond old social norms and create their own brave new world of cool (Yan, 2006). Indeed, it has been observed that Chinese youth seem to show even greater eagerness for individual expression than their counterparts in Western societies, as reflected in their online activities in comparison with their Western counterparts, as mentioned in Chapter 2. Nevertheless, it would be misleading to assume that Chinese youth today are totally Westernized, or they are exactly the same as their Western counterparts. As the aforementioned McKinsey study points out, at first glance, urban youth in China mirror their peers around the world: they embrace what's new and racy and walk around in branded clothes. But this first impression is deceptive, for a closer look reveals a consumer group that's a unique mix of modern tastes and traditional values, with behavior very different from that of its Western counterparts (St-Maurice and Wu, 2006). Likewise, it is observed that Chinese youth are undergoing a dramatic sex revolution since the reform started (Farrer, 2002; Yan, 2006), with pre-marital sex prevalent and frequent instances in which young people overtly challenge traditional sex and gender norms, as is illustrated by, for example, the cases of Mu Zimei and Sister Furong, the bloggers mentioned in Chapter 2. And yet, research meanwhile indicates that Chinese youth largely adhere to traditional gender roles and heterosexual norms (Liu, 2006a, 2006b; Wu et al., 2007; Lo, 2009). In short, it is not uncommon that Chinese youth draw upon historical figures and traditional values and lifestyles for inspiration when negotiating their 'modern' identities.

However, despite the cultural continuity, their eagerness for ku individuality suffices to form a striking generation gap between this generation and their parents, who grew up under dramatically different circumstances characterized by the Maoist Puritanism and collectivism. The two generations spent/are spending their formative years of youth in two strikingly different worlds: one was a closed society characterized by poverty, frugality, simplicity, ideology monopoly, lack of educational and information opportunities, whereas the other is an open society, characterized by globalization, prosperity, consumerism, pluralization of value systems and broadened educational and information accesses. This is why Michael Stanat (2005: 207), a teenager himself then, based on his own interaction with Shanghai teenagers, made the comment that 'China's intense modernization has produced a chasm between generations greater than any that has existed in the US.' Just as young people's expressive individualism with an emphasis on identity experiments and pleasure seeking often leads to parental disapproval, it has also caused widely shared social concern, which forms part of the reasons for which the younger generation was often criticized in the Chinese media until mid-2008, when during the Sichun earthquake its members displayed 'virtues' promoted by both traditional Chinese culture and today's subjectification regime, such as altruism, collectivism and heroism. This theme will be taken up further in Chapter 8, where I examine the post-80ers' collective narrative about themselves in an online forum. Now I shall turn to the other aspect of their individualism, which distinguishes them further from their Western counterparts.

THE PRAGMATIC GENERATION: CHINESE URBAN YOUTH AND UTILITARIAN INDIVIDUALISM

Just as individualism has impacted their lifestyle and tastes, it has also informed their life aspirations, resulting in an emphasis on what can be called 'entrepreneurial individuality', or 'economic/utilitarian individualism'. In line with this, today's young people predominantly and openly put individual economic success instead of collective interests as their priorities (Kwong, 1994; Liu, 2008a; Rosen, 2004; Yan, 2006). In a survey conducted in 2000 by the Communist Youth League and the National Student Federation on youth aspirations among 1,800 students in southern cities, the number one choice of these students was to be a billionaire; the second was to become a CEO of a multinational company. The students' third most popular ideal was to be a high-ranking cadre. This forms a sharp contrast with the youth of the 1980s, many of whom wanted to be China's Einstein or Mother Teresa (see Yan, 2006). In line with this trend, the urban youth interviewed in my previous study (Liu, 2006a) on the first-generation only-children young adults (born 1980–1983) were preoccupied with social-economic mobility through making money and gaining political power. In outlining their life ideals, they were strikingly clear and articulate about the kind of life they wished to live—a lifestyle which resembles, and even goes beyond, the middle-class life in Western developed societies. Houses (even villas with outdoor swimming pools), cars, extra money for children's education and modern leisure consumption, and travel are common features of such a life. Moreover, they insisted that this is the life everyone wants to live. This lends support to the observation made in an article on Chinese youth in *Asia Sentinel* that 'No other theme better marks this era than a fierce fight to get ahead, a relentless effort to achieve a better and more modern lifestyle' (Poon, 2008).

It is no wonder that, according to numerous surveys, Bill Gates has emerged as the most popular role model for Chinese youth (Yan, 2006) rather than the role models offered by the state such as Leifeng, Jiao Yulu and Liu Hulan, who sacrificed themselves, even their lives, for the revolutionary and collective cause. Interestingly, as Yan points out, it seems that they all know Bill Gates is the most successful entrepreneur and the richest man in the world, but few talk about the work of the Gates Foundation. Likewise, those who have achieved 'idol' status all represent material success and individuality at the same time. Take, for example, two of the most popular 'stars' in present-day China: Han Han, a writer and race car driver, who dropped out of high school and published a best-selling novel depicting the dark side of China's exam-oriented education system at the age of 17, and the 'Supergirl' contest winner Li Yuchun. The fact that Han sold more than 2 million copies of his books and could afford to have conspicuous consumption and to own five cars has made him the ultimate symbol of instant individual gratification for Chinese youth (Yan, 2006; Rosen, 2009). His rebellious attitude and witty posts have made his blog on sina.com the most accessed (Rosen, 2009). Li Yuchun, another popular youth icon, was chosen in an 'open' election to be the winner of the 'Supergirl' singing contest. Although criticized by establishment voices for her quirky personality, mediocre voice and androgynous appearance, these same attributes were considered positives or 'ku' by the more than 3.5 million, apparently, mostly young people, who voted for her (Rosen, 2009). And more importantly, she is admired also because her fame brings her material wealth as well. In a word, the two stars both embody economic success as well as individuality, both of which are deemed important by Chinese youth today.

All this means that youth today appropriate the double subjectification regime (see Chapter 1) in ways quite different from the Party state's expectation although it would be inaccurate to assume that the Chinese state has totally lost its ability to influence young people's minds (Lagerkvist, 2009; Rosen, 2009). In line with the ideal form of subject it wishes to produce, the Party state carefully tailored an ideal youth culture for youth consumption in the post-Mao era (Kwong, 1994). This pre-packaged youth culture entails double demands on young people in their identity formation in line with the double subjectification regime. Youth are expected to cultivate an entrepreneurial spirit necessary for national economic development and their own welfare (which seems especially necessary in a society where social security is underdeveloped), such as being diligent, intelligent, creative and knowledgeable. However, at the same time, they are to have the qualities once emphasized under Mao's socialism such as altruism, helpfulness and eagerness to work for the collective good. Moreover, they are to lead a simple life, listen to the Party and the government, and go wherever their skills are required (Kwong, 1994; Liu, 2006a). Obviously, presenting the ideal youth culture in this rarified form may solve the potential contradictions from a theoretical perspective, but it does not stand the test of youth's lived experiences of China's modernization, whereby socialist-collectivist idealism clashes head on with market competition, consumerism and individualism (Weber, 2002; Rosen, 2004, 2009; Yan, 2006; Kwong, 1994; K. Liu, 2004). However, this does not mean that the double set of enjoinments of the subjectification regime may not affect young people's identity work in any way. As we shall see throughout the subsequent chapters, trying to come to terms with conflicting and competing demands is part and parcel of Chinese youth's identity processes.

Rosen (2004) uses the expression 'the victory of materialism' to describe the value change among present-day Chinese youth. In contrast to urban young people of the 1980s, who were often 'searching for life's meaning', contemporary youth are success oriented and they openly seek 'the good life' based on materialism (Rosen, 2004; Xu, 2002). Seeking such a life, young people display an impressive pragmatism, which is reminiscent of the earlier-mentioned notion of utilitarian individualism (Bellah et al., 2008), featured with the long and painful act of deferred gratification and self-denial for the sake of economic success and social recognition. This seems highly paradoxical to their radical and romantic 'kuism'. Yan (2006: 257) provides a useful depiction of the type of pragmatism widely embraced among Chinese youth, observing that '[u]nlike their carefree and sometimes drifting counterparts in the United States, most Chinese teenagers and young adults are clearheaded about what they must do for a good life: earn high scores on exams, gain admission to a prestigious university, and secure a high-income job.' Yan further points out, drawing on existing survey statistics, that even younger members of this generation—those who normally would be more playful and less preoccupied with getting ahead—display similar trends toward dismal pragmatism. For instance, in 2005, the Research Center of Chinese Youth conducted a nationwide survey among students of primary and lower and upper secondary schools in both cities and the countryside. The results show that 57% of the students categorized their lives as unhappy because of the pressures of study, and more than 50% found the content of their textbooks boring or useless. Yet 58% of the students said they hope to earn a Ph.D., and 42% considered having good test scores the happiest thing in their lives. In comparison, only 19% regarded 'play as much as one wants' the happiest thing (see note 1).

By the same token, many Chinese youths tend to select subjects of study and jobs for the income they promise to bring in; but unlike their counterparts in the West, they rarely prioritize 'personal interest' in making their choice (Kwong, 1994; Liu, 2006a). In the aforementioned study of first-generation only-children young adults (Liu 2006a), almost all the college informants reported having 'chosen' their subjects of study according to rational calculations about the chance of getting enrolled at a university and the prospect of finding a 'good' job after graduation, which was defined as one that could bring lucrative incomes and social status, rather than one's interest. As one of the informants put it, 'All I have learned from the beginning of my life is that I can only do what I should do, not what I want to do.'

Chinese youth's practice of pragmatism further applies to the domain of political participation (Yan, 2006; Rosen, 2004, 2009; Liu, 2006a). It is widely observed that, as Yan (2006: 258) states, 'the current generation of Chinese youth is apolitical—indifferent toward both official ideology and the prospect of political reforms'. This is another feature that distinguishes them from their parents' generation, many of whom were 'Red Guards' actively participating in class struggle and from the young people of the 1980s, who gathered in Tiananmen Square demanding democracy and political freedom in 1989. This may have to do with the oppressive political culture that does not encourage political participation. However, more importantly, as indicated in Chapter 1 and Chapter 2, there is also evidence indicating that as the new middle class, who are a beneficiary of China's reforms and opening up, urban youth are largely reconciled with the political regime and take a pragmatic approach to politics (Damm, 2007). A strong interest in consumerism and identity politics has replaced the type of political enthusiasm of the pre-1989 generations (Damm, 2007). Interestingly, despite this pervasive political indifference, Chinese youth, especially the well-educated segment, have shown increasingly great interest in joining the Communist Party since the early 1990s (Rosen, 2004; Yan, 2006). A survey issued by the Organization Department of the Communist Party of China (CPC) Central Committee on the Party's 88th anniversary in 2009 showed more than 80% of new CPC members in 2008 were under 35 years of age, with members under 35 being the largest increasing age group since 2003. From 2002 to 2007, the number of college student Party members increased by more than 250% (*China Daily*, July 2, 2009). Moreover, more intriguingly, neither young applicants for Party membership nor young Party members care to conceal their real motivation to join the Party. Instead of indicating that they believe in the Party, many, although not all, openly acknowledge that they want the membership because it can enhance their chance for career promotion, as my study of only-children young adults shows (see Liu, 2006a), in agreement with Rosen's (2004) and

Yan's (2006) analyses. Likewise, university students show great interest in becoming public servants in governmental agencies, which as my informants in the aforementioned study (Liu, 2006a) explained, will lead to stable income and more importantly, will serve as the first step to become a leader with great power.

This lends support to Yan's (2006) argument that although Chinese youth, like their counterparts in the U.S., are individualists often radical in their attitudes, they do so only in the sphere of private life and mainly through personal consumption. For this reason, they have been referred to as 'rebels without a cause', who either as individuals or in groups explore few areas in the public sphere, with the widely observed popular nationalism among Chinese youth as a notable exception (Yan, 2006: 225). In Chapter 8, I will empirically examine how Chinese youth's normally 'apolitical' attitude and zealous nationalism together shape their negotiation of a 'Chinese' political self, as reflected by their online political (non)participation. The strong tendency of individualism, materialism and pragmatism among Chinese youth has led Yan (2006: 262) to conclude that today's Chinese youth lack idealism:

> If idealism, compassion, and naïveté are defining features of youthhood, the '80ers in China are hardly youthful. If efforts to resist and subvert the dominating ideology and symbolisms of the establishment are the core of youth culture, then China's youth hardly have a culture of their own.

Nevertheless, more recent observations indicate that present-day Chinese youth seem meanwhile desperate to believe in something and very willing to make sacrifices if they are persuaded that the cause is just (Rosen, 2009). The well-known popular nationalism among Chinese youth can be understood as an expression of this type of idealism. As Rosen (2009) argues, Chinese youth's nationalism 'represents a broader impulse and encompasses not only the defense of China against perceived enemies from abroad, but also the kind of love of country and self-sacrifice in support of those most in need' (361). The Sichuan earthquake presented an occasion to demonstrate their idealism, which was evident in youth's intense volunteerism following the earthquake. As Rosen (2009) further notes, this idealism is reminiscent of that of the Maoist period, as well as that during the 1989 Tiananmen Square demonstrations (Rosen, 2009). However, what is new and intriguing is that youth seem to be able to combine their nationalism with an admiration for foreign cultures and things, especially when these appeal to their 'ku' lifestyle. That is, they can display both tendencies at different times and even at the same time: participating in violent anti-Japanese protests while at the same time continuing to enjoy Japanese food, electronics and cartoons, showing patriotism by supporting the short-lived attempt to boycott Carrefour, the French superstore, in response to French interference with the 2008 Olympic torch relay during its Paris run and meanwhile making sure to use up all their discount coupons and finish their shopping prior to the May 1 boycott (Rosen, 2009).

CONCLUSION

In a way, the current generation of Chinese urban youth is the main beneficiary of China's reform and opening up in terms of material and economic prosperity, but it also bears the brunt of the transitional time (K. Liu, 2004). Reflective of the plurality of cultural influences in post-Mao China within the larger context of globalization, Chinese urban youth today display a complex and multi-faceted collective character that bears similarity to their Western counterparts, and yet is unique in itself. Thus, just as one can talk about modernization and the Internet with Chinese characteristics, as I have shown in the two preceding chapters, one can also discern an emerging urban youth culture with Chinese characteristics. As this chapter shows, the complexity and uniqueness is centrally reflected in that youth identity construction seems to revolve around an interrelated series of apparent paradoxes, or dualities. Most strikingly, Chinese urban youth appear to be both radicals eager for individual self-expressions *and* dismal pragmatists bent on the goal of the 'middle-class dream' based on material achievement and measured by certain pre-defined indicators, which apparently contradicts their eager pursuit of 'ku' individuality with its emphasis on individual differences. As such, it seems that they simultaneously embody both expressive individualism and economic or utilitarian individualism. This seems to indicate that whereas in the West these represent two major stages of the individualization process, in China they are occurring at the same time with equally forceful momentums. In other words, they seem to constitute two aspects of the same process, which is reflective of the rapid and dramatic social transformation China has been undergoing in the past three decades. Furthermore, and related to this, they are further characterized by a series of other paradoxical elements to varying extents: they are simultaneously materialistic and idealistic, internationalist and nationalist, global and local, apolitical but nationalistic, modern and traditional, blessed with material prosperity and yet under great pressure to strive for greater material wealth, and exposed to an unprecedented range of choices and yet lacking a sense of security.

No doubt such complexities of Chinese urban youth's identity reflect the material and cultural conditions of China's 'dual modernity' under 'socialism with Chinese characteristics', with its apparent contradictions, hybridity and non-synchronicity, as I have discussed in Chapter 1. Under such circumstances, youth have to form their modern identities at the intersection of tradition and globalization, and development and consumerism. In other words, this is a generation of youth that is torn between the traditional and modern, national and international, local and global influences and is in search of a direction. Much of the empirical chapters that will follow serve as illustrations of how Chinese urban youth try to come to terms with these seemingly contradictory and competing cultural influences through an examination of their experiences and perceptions of the Internet and their online self-representation. Chinese youth's 'dualism' has implications for the emerging 'modern' Chinese self, a point I will take up at the very end of the book.

The preliminary understanding enabled by this chapter of who Chinese urban youth are serves as a precondition for me to make sense of my informants' Internet-related narratives and Chinese youth's online self-representation in the rest of the book. As the book unfolds, we will gain a greater insight into the range of characteristics of their social-biographical situation covered here—but as experienced and defined by the young people themselves—that make the new medium of Internet relevant and meaningful (or otherwise) to them.

4 The Internet in the Everyday Lifeworld

'I-and-the-Internet' Narratives from Members of China's 'Net Generation'

Drawing on Schutz's (Schutz and Luckmann, 1973) notion of the everyday lifeworlds, this first empirical chapter explores how Chinese urban youth actively appropriate the Internet in ways that fit into their social-biographical situations and relevance structures. My exploration here draws upon the 70 semi-structured interviews with urban youth (aged 15–28) in a medium-sized city in North China during my fieldtrips in 2007 and 2008 (see the methodological note in the Introduction).

The 70 informants consisted of both men (32) and women (38). They were divided into three sub groups based on age: (1) 37 participants aged 15–18 (mostly high school students), (2) 21 participants aged 19–24 (including college students, vocational school students and non-student youth, some of whom were working and others of whom were unemployed at the time of my fieldwork) and (3) 12 participants aged 25–28 (all of whom were employed at the time and a few of whom were married). I shall pay attention to the differences in the participants' narratives that may emerge due to these structural factors. However, an in-depth analysis of these is beyond the scope of this book.

THE INTERNET IN THE EVERYDAY LIFEWORLD: A SCHUTZIAN PERSPECTIVE

The questions this chapter sets out to explore are fundamentally about Chinese urban youth's choice and meaning making in relation to the Internet. As I shall show in this section, Schutz's (Schutz and Luckmann, 1973) notion of the everyday lifeworld is useful to gain a 'situated' understanding of these questions in line with the general theoretical perspective adopted in this book.

Schutz (ibid.) defines the everyday lifeworld as 'man's fundamental and paramount reality' (3). It is 'the region of reality in which man [*sic*] can engage himself and which he can change while he operates in it by means of his animate organism' (ibid.: 3). The everyday lifeworld is the domain where one interacts with other people with whom he or she constructs a shared world. Everything one experiences within its limits is taken for granted as 'natural' or 'unproblematic until further notice' (ibid.: 4).

I find his notions of situation and relevance especially useful for this study. In every moment of conscious life, Schutz (ibid.) proposes, one finds oneself in a situation. Each situation involves the interplay between 'imposed' and 'open' elements (ibid.: 100–116), which constitutes the complex relationship between individual freedom and constraint (Bakardjieva, 2005).

What makes the term 'situation' relevant to my analysis here is that it not only can be used to designate the immediate micro-situation (e.g., I am sitting in front of the computer, writing this book right now), but it can also refer to the larger unit of time and space the individual inhabits at a given point or period of his life course (for an explication of the two types of situation, see Bakardjieva, 2005). Schutz (ibid.: 57) refers to the latter type of situation as 'the social-biographical situation of the individual with its involved complex of relevance structures, hierarchies of plans, and scopes of acts'. The meaning-bestowing categories defining such situations are formed intersubjectively and are basically imposed upon the individual and become interiorized by him or her (ibid.: 56–57). Such situations can be described in conventional social categories such as age, social-economic status, occupation, health status, family, social roles (e.g., gender roles) and so on. But, in line with Schutz's general notion of situation, one's social-biographical situation is meanwhile relatively open, and can be explicated without limit. At any moment of life, the individual finds him or herself in a specific situation characterized with a configuration of various categories (both imposed and open). Schutz (ibid.) mentions the instance of the individual's life stages to illustrate this point: 'The formal structures of childhood, youth, maturity, old age, etc., point to global social fluctuations in scope as well as great variations in content' (57).

I find the notion of 'social-biographical situation', namely the situation in a broader sense, especially useful for the purpose of this chapter. With this notion of 'situation' the individual's 'here and now' is expanded, linking one's self-defined situation to the specific socio-cultural context in which one's everyday lifeworld is embedded. That is, both 'the personal' and 'the social', and their complex and inseparable interaction, stand at the very core of this concept.

Schutz (ibid.) lays particular emphasis on the pragmatic motive that organizes action in each situation, insisting, 'The problems of action and choice must, therefore, have a central place in the analysis of the life-world' (18). The centrality of choice brings to the fore the notion of relevance: '[a]ll experiences and all acts are grounded on relevance structures. Every decision more or less explicitly introduces, besides the actor, a series of relevances' (ibid., 183). Relevance serves as the scheme of selection and interpretation concerning choices, decisions, and involvement of an individual (Schutz, 1970). It permits the individual to focus on one of—or the relation between—multiple 'virtual realms of reality, or finite provinces of meaning' in the lifeworld (Schutz, 1970: 7). In asking questions such as whether we see things in the right light, whether we should restrict our 'interest' to a given problem, whether something is of interest to me, whether something 'really matters' to me, and whether a matter is of consequence to me, we 'consciously inquire into our own relevance systems' (Schutz and Luckmann, 1973: 182). As such, relevance system (or structure), refers to the totality of those objects, features or concepts which are experienced as relevant in a particular situation (Schutz and Luckmann, 1973). Different individuals and groups embedded in their various situations have different relevance systems, depending on their hierarchies of plans, their knowledge as well as their general material and social environments. People grasp those aspects of reality that are of practical relevance, according to their unique position or situation (Schutz and Luckman, 1973). This logic equally applies to people's Internet-related choice and meaning making, as is illustrated by the case of the participants in this study.

Schutz (1970) indentifies various zones of interests, or relevances: the 'zone of primary relevance', the 'zone of minor relevance', the irrelevant zones, and the zone of 'absolute irrelevance' (111–112). He further emphasizes the constant changeability of relevance configurations as one moves from one situation to another. This notion of zones, or degrees, of relevance will be used in this study to explore the place or importance the Internet has acquired in my informants' everyday lifeworlds. Expressions such as interest, importance, usefulness, meaningfulness, impact, consequences, necessity/needs and so on, and their corresponding adjectives (and the opposite of these) will be important for describing the young people's relationship with the Net.

Schutz (1970) also distinguishes between 'imposed' relevances and 'intrinsic' or 'volitional' relevances, which in daily life are often intermingled with one another and seldom found in a pure state (113). The former types of relevances are those imposed on us by events and situations, while the latter types are the outcome of our spontaneously chosen interests and goals (Schutz, 1970: 114). Although it will be difficult to establish unequivocally 'imposed' and 'intrinsic' relevances in many cases, I still see in Schutz's notion of intrinsic and imposed relevance an analytical tool that could be useful in the investigation of how the tension between the competing discourses of the Internet are being played out in the young people's everyday lives embedded in the socio-cultural context of today's China. The construct of the computer as an educational device in line with the discourse of the information society, is being constantly challenged by the alternative construction of the computer as an entertainment and leisure device (Facer et al., 2001; Sandvig, 2006). The socially approved uses of the Net (i.e., those for educational purposes) often stand in contrast with what children and young people actually use it for. Children and young people are far more fascinated by the Internet as a communication and entertainment medium than as a learning and information resource (Buckingham, 2003; Burn and Willett, 2005; Facer et al., 2001; Golub and Lingley, 2008; Livingstone, 2003; Livingstone and Bober, 2004; Sandvig, 2006; Seiter, 2005). Viewed from the Schutzian perspective, uses for educational purposes constitute 'imposed relevance' for children and young people, whereas uses for recreational and social purposes are more in line with their 'intrinsic relevance/interest'. One important question to explore is how the urban youth in this study are trying to come to terms with the tension between the two types of relevances of the Internet.

To sum up, Schutz's notions of the everyday lifeworld, situation and relevance enable me to view the participants as a socially situated (albeit active) individuals and interpret their Internet-related choice, decisions and preferences in relation to the larger pictures of their lives embedded in specific socio-cultural and institutional contexts, with a variety of situational characteristics, as experienced and defined by the subjects themselves.

'IT'S JUST ENTERTAINMENT.': THE INTERNET AND RECREATION

Generally speaking, users throughout the world employ the Internet in multiple ways: socially, instrumentally and recreationally. However, from a Schutzian perspective, individuals in their specific social-biographical situations with their specific relevance structures appropriate the Internet in ways that make sense for their own purposes. Thus, if we conceptualize Internet use as a spectrum with these different use purposes (social, recreational and instrumental) across it, then we could locate individuals' or groups' trajectories across the surface.

By asking the informants about their everyday online activities, I got to know that most of the informants (64 out of 70) used the Net mainly for recreation—a term they used interchangeably with 'entertainment', 'pastime', 'playing' and 'fun' (for the purpose of convenience, I shall also use these interchangeably in this book). This seems to reflect the general Chinese online culture, which tends to be predominantly entertainment oriented, as I show in Chapter 2. When asked how they would define the Internet, a predominant reply I got was 'just entertainment'. That is, they simply drew an equation mark between the Net and entertainment. Then, it is not surprising that many of them used the verb 'play' when talking about the Internet, such as 'play with the computer', 'play online' or 'my mother does not allow me to play [online]'. A few of them even referred to themselves as 'players' (*wanjia*). For example, Liaolong (an 18-year-old high school student) and his two friends at a group interview told me that they were merely *cainiaoer* (literally 'birds for food', which means green hands) compared with those really great *wanjia* ('players'), although they had played all kinds of games and come to the top in each of them, as they explained to me.

Only a few (6, mostly those aged 25–28) reported using the Net mainly for instrumental purposes, such as things related to work and study. Although they were the minority (6 versus 64), these users constituted a marked contrast with the earlier-mentioned participants who employed the Internet almost purely for 'playing'. As such, two distinctive camps of users emerged from the informants' narratives. Markham (1998: 86–87) in her book *Life Online: Researching Real Experience in Virtual Space* discerns three types of user experience of the new medium: the Internet as a 'tool', as a place and as way of being. According to Bakardjieva (2005: 103–4), the 'tool' and the 'place' and 'way of being' perceptions respectively correspond to the 'instrumental' and 'intimate/expressive' user-technology relationships observed by Turkle (1984) and Aune (1996) in their studies of user-computer relationships. Admittedly, such a dichotomy can be questioned in many cases, in which the different types of user experiences are best seen as a continuum instead of a dichotomy. However, the two ideal types, 'expressive' and 'instrumental' users, do seem to be useful in gaining an understanding of the contrasting Internet perceptions and experiences of the youth in this study. Such a contrast deserves an in-depth exploration, which will be a task for the next chapter, in which I will present in further detail the 'instrumental' users. Here in this chapter I will focus on those who reported using the Internet mainly or purely for entertainment.

A variety of online recreations were named by the informants, ranging from playing online games to watching movies, listening to music, reading novels, chatting, reading and writing blogs and decorating one's online personal space or visiting other people's personal spaces. The online activities could vary from person to person. But a gendered pattern is often, though not always, noticeable, with more boys playing games, and more girls engaging in other forms of online recreation and boys showing greater interest in and self-confidence about the new technology than girls. This is in line with findings in some other studies on both Chinese and Western youth (e.g., Li and Kirkup, 2007; Lin and Yu, 2008; Seiter, 2005). Moreover, the frequency and duration of being online may also vary according to the contingencies of one's social-biographical situation such as one's amount of unstructured time and life goals, which is often (but not always) related to one's age and status as student or non-student. However, for them, the predominant purpose of going online is entertainment. As I shall show a bit later on, the social-communicative affordances of the Net, such as chatting, are also employed mainly for 'pastime'. Most of the vocational school students and non-student young adults who were not working at the time reported that their online time varied from 5–10 hours per day, but purely for entertainment (in this volume I refer to them as the 'heavy recreational users'). Most of the middle school students and college students in this study reported going online only in their 'free' time, such as weekends, lunch breaks, summer and winter vacations after exams, amounting to 2–6 hours a week, but almost solely for recreation, or as they put it, 'just to relax' (I refer to these as the 'moderate recreational users'). And the *wangba* (Internet café) was the place where most of their online time was spent, although most of them did go online at home as well whenever they had a chance. Some of the heavy users reported that they stayed at the *wangba* or to use their words, 'steeped themselves in the *wangba*' (*pao wangba*), all day long every day (sometimes even overnight) except for extreme cases such as when one is sick. However, this differed for the middle school students, who typically did not have much free time; as one middle school male student told me, 'if we do not have more than two hours, we will not go to the *wangba* because one cannot do (he meant 'play') anything within less than two hours'. In Chapter 5, I will explore what *wangba* going means for the goers from their own point of view.

It is, however, noteworthy that although my informants emphasized the recreational function of the Internet, this does not mean they failed to see the multiple affordances of the new technology. For example, although most of them conceptualized the Internet as 'just recreation', they meanwhile recognized its function as a window into the outside world and as a (at least potentially) helpful information source. This is true especially in the case of the young adults. For instance, some of them told me they sometimes used the Internet to look up information before they traveled. Many occasionally did e-shopping and found it cheaper and easier as well as offering greater varieties of goods, as 'on the Net one is selecting from a national rather than the local pool with a limited choice', as a young man (22-year-old, unemployed) told me, showing me his leather wallet newly bought on the Net. Another young man (24-year-old, unemployed) told me he once bought an Adidas sport suit online because one could not find the same style and quality in that city. A female student at a vocational school told me that she and her boyfriend had been managing an online shop 'selling "ku" things that appeal to young people'. And they often browsed the Net to learn from others who manage businesses online. As such, the young people did employ the Net as a means to satisfy their consumption needs in line with their modern 'ku' taste described in Chapter 3. Nevertheless, such uses were rare, my informants emphasized, as they used the Internet mainly for recreation.

Tracing their trajectory across the aforementioned conceptual spectrum, I found that my informants rarely used the Internet for instrumental purposes, such as studying, information seeking, news retrieval and so on. With regard to the student informants, I was especially struck by the rarity of their use of the Internet in relation to school work. This irrelevance of the Internet to academic work constitutes a strong theme which will be examined in a separate section.

This particular kind of user-Internet relationship is to be played out more fully in the rest of this chapter, where we shall see how, from their own perspective, their predominantly recreational orientation toward the Net was related to their social-biographical situations and relevance systems and how this shaped the role of the Net in some other domains of their lives.

'CHATTING IS CHATTING.': THE INTERNET AND SOCIALITY

The Internet has been much celebrated for its potential to facilitate social-communicative activities and community building regardless of geographical boundaries. It is reasonable to assume that this affordance may be especially cherished by only-children such as those in this study, given the lack of siblings. However, in the case of my informants, it would be exaggerating to imagine these young people desperately reaching out to other people far and near. As I shall show next, although they did value the Internet as a social-communicative means, their circle of online friends were mainly from the local community. Neither did the relevance of the Internet as a communication means override its importance as an entertainment device. Rather, their strong preoccupation with online recreation spilled into their online communication.

I learned during the interviews that the majority of them did not have an e-mail account and many of them had not even heard of the term 'e-mail'. The few who had an account indicated that it had been seldom used. Their online communication, in case of need, was mainly through chatting in QQ, a site run by a company named Tencent which offers the IRC function and a MySpace-like social networking site (where one can do blogging, among other things). Although QQ has made it possible for the user to get in touch with people all over the country, or even beyond, the contacts included in these young people's lists were mainly local friends, family members, former classmates and friends and relatives who were not living nearby. Few of them reported having participated in bulletin board system (BBS) discussions. When asked if they had not made use of the Net to make friends all over the world or to reach out into the outside world, they said that although they had tried to talk with strangers online, that was mainly when the Internet was first introduced into their lives and they did that out of curiosity and mainly when they could not find other things to do. Most of them emphasized that their online socializing with 'strangers' was mainly limited to their own 'groups' or 'tribes' ('*jiazu*') within games, where people within the game camp up as one group against another group (or other groups). Although in some rare cases some had made great friends from the other side of the 'screen', or had even met friends from the virtual world in real life, generally speaking, the young people deliberately made a distinction between 'real life friends' and 'friends in the virtual world', a point I will return to later on in this chapter. Some of them emphasized that online socializing and subsequent meeting between *wangyou* (Net friends) was popular only when the Internet was first popularized. Today, they indicated, most young people find meeting *wangyou* naïve and unrealistic as well as unnecessary.

This does not mean that these young people, who are mostly only-children, did not value the Internet as a means of communication and socializing. It simply had to do with how their everyday lives were structured. As Zhouchao (21-year-old, male vocational school student) said: 'We have enough friends around. Our communication is basically with people around us. So there is no such need to make friends at a distance'. Or as Wangchen (18-year-old girl) explained, offering a shared understanding among the informants, 'Our daily life is largely organized around school and the local society. Nearly all our friends and families are nearby. There is no such need. The Internet for us is mainly for relaxation'. This remark brings us back to their everyday life actualities with their own relevance systems and, related to this, to their preoccupation with online entertainment.

Moreover, such a choice also has to do with the fact that many of the informants found it meaningless and boring to chat with strangers. For example, Gaofei (20-year-old, female vocational school student) told me:

> It is very boring. Always like, when we meet (online), we ask 'what is your name, where are you from, how old are you' and so on. How boring! It is seldom that the chat goes well. . . . Even though you can chat deeply, you are not sure if the other side is deceiving you or not. Even in a rare case where we do meet Net friends, most are from the same city.

As is clear from this example, the sense of meaninglessness also had to do with their observation that it is hard to find out the truth about the other side. Exactly due to the blurred boundary between 'truth' and 'deception' in computer-mediated communication, it was perceived as a risky business to engage in 'Net romance' (*wanglian*). As Net romance has been much covered in the media as a widely existing phenomenon in China (Xinhua News Agency, December 29, 2007), I was curious whether these young people had experienced '*wanglian*' to find a girlfriend or boyfriend. Although they recognized that it is not impossible for true love to develop online, most of them indicated that *wanglian* is not realistic, citing the saying '*jianguang si*' (once the lovers walk off the Net into the light, or once they meet in the 'real' life, the love dies). Moreover, and more importantly, further invoking the notion that one's use of the Net is framed by 'real life' contingencies, a 22-year-old woman vocational school student insisted that it is only those who do not have enough self-confidence in the 'real' life that turn to the Net to look for 'admirers'. Although such assumptions can be challenged, they do invoke the notion of 'relevance' and 'need' in framing individual choice. Furthermore, their 'realistic' approach to 'net-based' intimate relationships also seems to have to do with one's only-child status, a significant feature of their social-biographical situations. Some of my informants indicated that as an only-child one is expected by the parents to live close by in the future, which is in line with the findings in an earlier study I conducted on the only-child generation and intergenerational relationships (Liu, 2008b). This, together with the shared understanding that it is beneficial for both when the young couple are from the same city, makes it more likely that young people try to find a partner within the same city. And sometimes, online communication does help one with this goal. For example, Jianglong (22-year-old, vocational school student) met his girlfriend in a chat room and found out that she was also from the same city, even the same school. After some rounds of chatting for a while, they appointed to meet. He told me that it was exactly because they were both from the same city that they found the relationship more feasible and worth pursuing.

Although most of them tried to avoid more serious emotional engagement with online 'friends', they did employ the social networking sites, especially online chatting to seek fun. I heard two opposite reports concerning online chatting. Most of the informants indicated that they seldom chatted. Or as Zhanglin (17-year-old student) put it, representing his friends at the same interview, 'We chat only when we cannot find other things to do online,' which was a shared experience among most of the interviewees, both male and female across the two age groups. They explained that almost all their friends and relatives were nearby and they could meet regularly, and so it was not necessary to chat online. Moreover, they explained, as everyone has a mobile phone, they sent text messages to each other, or just made calls. There were, however, a few others who reported that each time they were online, they would chat with friends. And the purpose, as Fanxiang (17-year-old) explained to me when I was interviewing him and some classmates, was to 'crack jokes with friends and pass on messages'. Thus, the two schools of 'chatters' shared a tendency to use online communication for leisure purposes, including seeking fun.

To demonstrate their tendency to seek fun through online communication, it was typical that friends played tricks on each other by pretending to be someone (usually a stranger) from the other sex, flirting with the person (usually from the same sex) he or she wished to fool for fun. For these young people any day could be turned into an 'April Fool's Day'. This was possible because of the anonymity characteristic of online communication. Almost every one of them had more than one QQ number. One was shared among friends while the other(s) kept a secret for such purposes, or they could simply borrow a QQ number from an acquaintance, so that the person one struck a chat with would not recognize who it was. Moreover, such jokes were effective also because of the heterosexual norms in Chinese cyberspace (Wu et al., 2007; Lo, 2009), which reflects and reproduces the offline gender culture in the

Chinese society (Croll, 1995; Liu, 2006b, 2006c). They told me that their online chatting almost always took place between a female and a male person. Some of them explained this to me as follows:

> It is just natural. The opposite sexes attract each other. One usually finds much to say with someone of the other sex, whereas if the other side is of the same sex, it is hard to find a topic.

A girl at the same interview added: 'People would think there is something wrong with you if you chat with someone of your own sex'.

This reflects the strong heterosexual norms and the still widespread homophobia in today's China. Furthermore, it may represent a type of compensation for the inadequate space in the 'real' life for cross-sex friendship, given the widely observed sex-segregation in Chinese schools and beyond (Liu, 2006c).

In short, as a few of the informants emphasized, 'chatting is chatting'; it is not for anything serious, but for fun. A 22-year-old man, taxi driver, even used the term 'play chatting' when he was telling me about his online activities. Such a view may neglect the fact that chatting does sometimes handle more serious topics, and yet, it does reflect the type of recreational orientation toward the Net widely found among the Chinese youth. This tendency was further observed in their use of the 'my space' (*wo de kongjian*) at the social networking site QQ, which, according to some of my informants, mainly served as a space for reading and circulating jokes rather than for blogging and communicating with others. For example, Wangsi (16-year-old) told me that her 'my space' is purely used for collecting humorous stories, which give her a hearty laugh from time to time when she is tired from studying and when she is not happy.

In many ways, the Chinese youth in this study employed the Internet for social-communicative purposes in similar manners to their Western counterparts. For example, just as youth in Norway and some other Western societies, such as the US, are much attracted to social networking sites such as Facebook, Friendster and MySpace, the majority of urban teenagers in China have an account at the aforementioned QQ. Besides, they seem to employ the Internet for sociality in ways that lend support to research findings done in the West showing that in most cases the Internet strengthens young people's local networks rather than extending their networks beyond, although it has that potential and although in some cases some have made friends and built community with people at a distance (Boyd, 2008; Ellison et al., 2007; Gross et al., 2002; Livingstone, 2002; Livingstone and Bober, 2005; Wellman and Gulia, 1999). Furthermore, they mostly used the mobile phone, especially text messages (for more on the popularity of short message service and text communication in China see Latham, 2007) to contact friends and relatives. Again, this preference for the mobile phone as the major communication method is in line with research findings made in other societies (e.g., Livingstone and Bober, 2005; Liu, 2010).

However, it seems to me that generally speaking, compared with their Western counterparts, the Chinese youth in this study seemed to show a greater interest in combining online communication with entertainment, or a greater degree of 'playfulness' in their online social-communicative activities. This reflects their preoccupation with online entertainment, which had directly to do with their perceived needs of the self situated in a Chinese socio-cultural context.

'WHEN WE ARE FRUSTRATED IN REAL LIFE, WE TURN TO THE NET.': THE INTERNET AND THE SELF

When asked if they could come up with a metaphor or metaphors to describe the Internet, many of my informants referred readily to the Internet as just a 'virtual world' (*xuhuan shijie*), in contrast to the 'real world'. In other words, they tended to make a clear distinction between the 'two' worlds. They further insisted that it is exactly the 'virtual' nature of the Net that was interesting to them. Or as Liwei (17-year-old boy) argued, 'If it were the same as reality, it would not be able to attract people.' Similarly, as already mentioned, some of the participants deliberately made a distinction between 'real life friends' and 'friends in the virtual world', who serve different functions. For example, a 16-year-old student claimed that the friends he interacted with in the games 'are more generous and more like those ideal friends in traditional legends who were willing to die for you'. But, he added, 'once we meet in real life, they would be the same as other people.' Thus, it was the virtual nature of the figures that seemed more interesting to them, and therefore, they preferred to keep the online world online, or virtual, rather than to destroy its more 'ideal' form.

However, from the theoretical perspective adopted in this study, this does not mean that the offline and online lives are isolated from each other. Rather, they are closely interrelated. This perspective coincides with my informants' 'I-and-the-Internet' narratives. Accounting for their particular orientation toward the Net, my informants actively struck a link between the 'real' and 'virtual' world, making associations and comparisons between the 'two' worlds, although they constantly saw the two as two parallel, but different, arenas of action and interaction. In this process, some of the themes regarding Chinese urban youth discussed in Chapter 3 were cited as salient factors which had directly shaped how they related to the Net. Being 'Chinese', 'only-children', 'situated in a small city' and 'young' were seen as the most important factors of their socio-biographical situations that framed their particularly 'playful' relationship with the Net.

In explaining why they were so preoccupied with online recreation and why the 'virtual' world, or more exactly the 'virtual' part of cyberspace, was of particular interest to them, a shared explanation was that the pressure (*jali*) on the self in 'real' life was too great. This voice was particularly resonant among the middle school students, echoed by the older participants recalling their middle school days. Reflecting the general social-biographical situation of Chinese urban youth (see Chapter 3), they indicated that the pressure mainly came from high parental expectations typical of the only-child family, the exam-oriented educational system and fierce competition in society and the changed standard of a 'good life'. This forms a strong basis upon which my informants compared the Chinese case with the Western countries where they believed things are quite different. Such comparisons, which reflect the widely observed tendency among Chinese youth to hold the Western developed societies as the ideal 'other', appeared time and again in my informants' narratives. Here are some of the typical complaints they uttered:

> We have too much homework, too many exams! We are under constant evaluation and are being ranked according to the test results all the time. Besides, we never seem good enough in our parents' eyes. (Qihui, 16-year-old girl)

> Parents and teachers keep telling us that the competition in the future will be extremely fierce. One must strive hard in order to have a standing point in society (*yixi zhidi*). (Zhanglei, 17-year-old girl)

> Also for our parents, so they will be living a good life in the future. So everyone wishes to enter a good university. We are under too great pressure. (Liuchang, 16-year-old girl)

> School life is too boring and tiring. It's almost all about studying, nothing else. Not like in the West. As far as I know, few adolescents there play online games. It is mainly people who are 23–24 or above that do that. But in China, it is mainly those at 17–18, or 16–17, high school students like us who play games. So I think it has to do with the education system. We are under too much pressure. (Jiantao, 17-year-old boy)

'When we are frustrated in real life,' emphasized Wangli (17-year-old, girl) at a group interview, 'we turn to the Net'. Another informant (16-year-old boy) told me, 'When I do very badly in an exam, I would just immerse myself in the virtual world, playing and playing, not caring about anything in the 'real' world.' The following conversation further highlights the major difference between the 'real' world and the 'virtual world of games' as perceived by the Chinese youth:

> *Zhufeng (17-year-old, boy):* The greatest difference between the two is that the virtual world is easier than the real world. In reality, you have to struggle many years to upgrade yourself. But in the virtual, you can do that by merely overcoming a few tasks.

Cuiqian (17-year-old, boy): Besides, in real life you cannot foresee the future, but in games, you can. There you know what is going to happen. You then have a sense of being able to foresee things.

Wangfeng (17-year-old, boy): And each time you have obtained good equipment, you feel that you are so powerful. People will admire you. It is just to satisfy your sense of vanity.

Although most of the older informants were no longer under the same pressure for the College Entrance Exam, either because they had entered college or because they had given up the academic pursuit, the pressure to strive forward and achieve was equally real for many of them. This was true particularly given the difficulty for college graduates and young people in general in finding a 'good' job in the competitive labor market and/or to make money in order to secure a 'good' life for themselves and their families in a society where what counts as a 'good' life has been redefined and where there is a general lack of social security, as I have indicated in Chapter 3. Under such circumstances, like the younger informants, they also found it easier to make achievement and prove their power in games. Just as Machao (23-year-old, unemployed) remarked, 'In games, one can upgrade oneself as long as one spends enough money and time. I have a sense of fulfilment that is hard to get in real life. It satisfies one's sense of vanity.'

The sense of achievement and being in control may be especially relevant for boys and men, who are supposed to achieve something in life according to the traditional gendered expectations still very prevalent in today's China (Liu, 2006b, 2006c). However, as research indicates, urban girls and young women in China are equally pressured to achieve due to interrelated factors such as only-child status, high parental expectations regardless of the child's sex and their role as 'substitute sons' (Croll,1995; Liu, 2006a, 2006b). Thus, although they tended not to play online games so much, especially the kind enjoyed by boys, my female informants also showed great appreciation of the 'virtual' world, where one 'does not need to think about school and parents and all the things in *this* world', as Zhaofei (17-year-old girl) told me.

In addition to the pressure for achievement, and related to it, lacking space for individual expression was cited by the informants as a main reason why they needed to escape to the 'virtual world'. It was a largely shared perception among both the middle school students and the young adults that their parents and grandparents could not understand them and they indicated a general lack of space for individual expression. This might be a typical remark one expects to hear from most teenagers and other young people. Nevertheless, the participants in this study could ground their argument in their lived experiences as urban only-children in the specific context of today's China. Sunhao (17-year-old, high school student) said on behalf of his friends at a group interview, 'It is the same for all of us only-children. Our parents have planned each and every step of our life course according to *their* beliefs. And we are to go according to that plan.' Again, I heard a similar comment from some of the young adults, who typically combined their complaint with an admiration for the West. For example, here is an interview excerpt of two friends (female vocational school students), explaining why they were fascinated by the online world:

> *A*: They (parents) have a kind of thinking mode. And you must follow theirs. That is, they design a human life future for you; you then must go long that road. They do not care if this road is suitable for you or not. And you cannot try out anything without their permission. They protect, or rather, control you all the time.
>
> *B:* In Western countries, once one becomes 18, one can decide for oneself. Parents do not tell you what to do any longer.

According to the informants, one enjoys greater freedom in the 'virtual world' in contrast with the 'real' life. That is, they claimed, uttering a utopian view of the liberating Internet, online one can be whoever one wants to be and do whatever one wants to do without the same consequences as in 'real' life, thus making up for what is lacking or impossible in the 'real' life. Or in Zhou-wei's (21-year-old, woman, vocational school student) words, 'In the virtual world of the Net, *everyone* can have another self as *everyone* has something he or she is not satisfied with in real life.' As such, online venues such as 'my space' (*wo de kongjian)* offered by social networking sites such as QQ and games became popular spaces for self-expression and identity experimentation often not available in the 'real' life. For instance, Zhangmi (22-year-old, woman, vocational school student) argued, 'One can dance in online dancing games in ways not possible in the 'real' life.' Similarly, some of them explained to me that one can 'get married' very easily in online games with whoever one happens to like without taking the same consequences as one would in the 'real' life (see Chapter 2 for a short overview of Chinese youth and 'web marriage'). And indeed, many of my informants reported 'being married' with someone/some people on the other side of the screen and having 'married' and 'divorced' many times. 'It is just for fun,' emphasized Linlin, a 22-year-old woman college student.

The manners in which one can employ the 'freedom' allowed for by the Net for self-expression are not limited to just these examples. Many of my informants made use of the 'my space' as a venue to 'display one's self' (*zhanxian ziwo*) in their words. A shared purpose among those who invest much time and money in their online 'my space' was to display 'individuality' (*ge xing*). Such 'displays' were often done through consumption of 'virtual goods' just as it often depends on consumer choice in offline life. Some of my informants told me that they often bought virtual items to change their avatars' appearances and to decorate their 'my space'. They also bought virtual items for their friends' birthday presents. One of the female young adults, a 21-year-old vocational school student, lamented that she really wished she could buy the virtual vest which she found extremely 'ku' for her boyfriend, but it was too expensive. When asked why she preferred to buy 'virtual' instead of 'real' things for him, she insisted that 'it is impossible to find such special things in real life.'

Their uses of 'my space' varied from case to case, ranging from representing one's imagined future home with modern material comfort to expressing feelings for their boyfriends or girlfriends, collecting and sharing one's favorite music, photos, prose, poems, proverbs, cartoons, humorous stories and jokes and letting out bad feelings. To illustrate the last type of use, Zhaoyi (23-year-old, new graduate from a vocational school) offered to show me her 'my space', which she had used much time and a considerable amount of money to make and re-make. She reported having spent 100–200 yuan every month, most of her pocket money, since five years before on 'virtual items' for her 'my space'. She explained that she chose darkness, blood and death as the dominating themes of her space because she felt very depressed in the 'real' life. Whereas she must hide such sentiments in the 'real' life, she found a 'free space' here to express it. Having a 'fancy' 'my space' displaying individuality as well as computer/Internet skills also provided a chance for one to gain admiration for one's technical, hence 'modern', competence and 'ku' aesthetic taste from peers who visited each others' 'my space'.

In a word, the young people portrayed the online world as the ideal 'other' of the 'real' world, albeit the self in the 'virtual' world is directly related to their 'real' life experiences, serving as an extension and compensation of the latter. For them, whereas the former is relaxing, exciting, cheaper, easier, beautiful and liberating, the latter represents the opposite. Given such perceptions, it is not surprising that many young people tended to escape into the virtual world as a strategic means to counter the pressure and/or lack of individual space in the 'real' life.

However, in the case of those who often spent many hours a day at the Net café, predominantly the non-student young adults (often unemployed) and the vocational school students, this is only part of the story. Besides making a link between the lack of space for individual expression and their resort to the Net, these young adults reported that they 'steeped themselves in the Net café' *just* to kill time because in 'real' life, they did not have anything else to do and so they felt very bored with life. Lixue (female, 22-year-old, vocational school students) spoke for her friends in a group interview, 'We have nothing else to do. Time is the most we have of.'

Some of the vocational school students attributed this 'nothing else to do' to their personality, claiming that 'we are not the type that like studying'. Others found some institutional reasons for their relationship with the Net. For example, Wangjin (22-year-old female vocational school student) explained, 'We usually do not have many classes. And it is very boring at school. And there is only a Net café nearby, so we just go there.' Thus, she attributed their having 'nothing else to do' to having too much unstructured time at school. Interestingly, however, the opposite, namely, a much structured school life, may also lead to a similar perception. For instance, according to Zhaoli (21-year-old, female vocational school student):

> The reason for Chinese young people to indulge themselves in online entertainment is that we are bored with life (*wuliao*). Not like in the West, where students can work part-time (*dagong*), our daily time is occupied by having classes. Things we can do are really too limited.

Another vocational school student at the same interview strongly agreed with Zhaoli by adding, 'Indeed. The things we can do are *really* too limited. Our city is a small one. Unlike in big cities, where there are more opportunities for part-time jobs and many interesting things to do. We are *really* limited here.' She told me she would move to a bigger city upon graduation as her uncle would find her something to do there.

Thus, my informants saw young people's boredom with life as a uniquely Chinese phenomenon, especially in smaller places like their home city, although similar situations may be true of Western youth and children, who also often resort to the mass media for alternatives (Boyd, 2008; Livingstone, 2002; Buckingham, 2006).

Still others attributed their idleness to the fact that their family had arranged everything for them so that they were not under much pressure to work hard, which is somehow paradoxical to the observation that Chinese youngsters are under great pressure to achieve. Zhao and Zhang, two young women friends from the same class at a vocational school, responded to my question 'Do you have nothing to do? Are you not studying?' as follows:

> *Zhao:* We are not good students. Good students seldom go online.
>
> *Zhang:* Yes. Good students also go online. But they do not do that as much as we do. They want to, but they are pressured to study. Unlike them, we do not need to worry about anything. Our families have arranged jobs and everything for us.
>
> *Zhao:* That's true. Our families have arranged it well for us.

This shows that family social economic status may have a role in shaping how one employs the Net. This is further supported by the case of Liwei and Hetian. The former was a 24-year-old junior middle school graduate, who had spent the years since he left school mainly playing online games and who once spent 30,000 yuan within a month just to play online games. That he could afford to immerse himself in the Net café was not only because, as he claimed, he is a person who is naturally free from any care and worry, but also because his father was a chief labor contractor (*baogong tou*), which in China implies high private profits. Hetian (employee at the local grain bureau), at the age of 27, was still indulging himself in online games and frequently patronizing the Net café. This forms a contrast with some of the other older participants in this study, who seemed to have either nearly stopped using the Internet or just used it for instrumental purposes, as I shall show in Chapter 5. Although he found it dispiriting and wished to correct himself, it had proved hard to do so when he was not under any pressure to make a living on his own, being from one of the richest families in China.

What is striking is that for many of these young adults, it was not even interesting to 'hang oneself on the Net', to use their expression, so many hours every day. When asked, 'why do you do that then?', the reply was again that because they had nothing else to do, or, as Wanliang (24-year-old, male, non-student, unemployed) said, 'because we are used to it. It has become a habit.' Thus, it seems that for them, 'hanging oneself on the Net' had become a life style or a mode of being.

The shared experience of 'nothing else to do' among some of the young adults points to the paradoxical situation typical of the urban only-child generation. On the one hand, this generation carries great parental expectations and social pressure to achieve for the sake of one's own and the family's future welfare so that they constantly feel the need to strive hard. On the other hand, as the main body of China's emerging middle-class, most urban only-child families can afford for their child's 'emerging adulthood' (Nelson and Chen, 2007). Being on the Net serves as an important means for young adults to kill the time allowed for by such a 'privilege'. Nevertheless, it seems that this kind of leisureliness is not always experienced as a pleasure. Rather, it goes hand in hand with uncertainty, frustration and boredom. Moreover, the perceived 'nothing else to do' also represents urban youth's 'choice' in a socio-cultural context characterized by a lack of social and physical space and limited range of available activities for young people (Giese, 2006), a point I will come back to in Chapter 6.

As we have seen so far, the young people's relationships with the Net were shaped by multiple factors in their 'real' lives. However, given the interrelatedness of online and offline life, the effect was never a one-way train (Fletcher, 1998). According to my informants, their online life did impact their 'real' life self in multiple ways, which in the case of the heavy users was often perceived as largely negative. And this was not limited to the perceived harm over-indulgence in the online world may do to one's health, neither to the shared perception that indulging oneself in the 'virtual' entertainment does affect one's study and by extension, work or other 'proper' businesses in life (a theme I shall discuss in further detail in the following section). Moreover, and perhaps related to this, as many of them pointed out, no matter how one indulges oneself in the 'virtual' world, one has to come back to the 'real' life. Coming back from the virtual, the discrepancy between the 'two' worlds often makes one feel that 'real' life appears even harder. Thus, Xiaohan (17-year-old, middle school student) reported that sometimes after playing online games he had a sense of achievement there, whereas he felt lost and empty in real life, where the real and harder tasks were waiting for him. Then it may end up in a vicious circle, as Yutian (24, male, employed) a heavy recreational user, concluded talking about his perception of the dialectic between the 'real' and the 'virtual' world, 'The more empty (*kongxu*) I feel, the more I play, but the more I play, the more empty I feel.'

Furthermore, some of my informants experienced the Net as having the effect of alienating the self, just as some researchers have argued (e.g., Kraut et al., 1998; Nie and Erbring, 2000):

> People who are often online do not wish to be seen by people in real life as they are so *weimi buzhen* (dispirited). They like to be on the Net, where they have a sense of being able and powerful. On the net, they can use their 'shell' (*quke*) to hide themselves. No one finds out. When I play often, I tend to go without washing and grooming myself, hair uncombed and beard unbarbered. Gradually, it becomes a habit, and then I do not like to be together with other people, but like to meet people online. I feel I cannot cheer up (get spirited) in real life. If I had not been online so much, I would have appeared much smarter *(jingshen)* than now. (Gulin, 23-year-old, male young adult, unemployed)

Besides the previously mentioned, one's 'virtual' life experiences can permeate the 'real' life in other ways as well. For example, some heavy gamers reported that sometimes it is hard to stop 'living' in the virtual world, even though one's body has come back to the 'real'. Thus, Zhaohui (21-year-old, vocational school student), who liked playing a game in which she was an angel, told me she bought a new pair of wings for herself in the game and practiced walking with the wings before she could take off flying. When she was walking in the street in the 'real' life, she tended to walk in the same way practicing to walk with her new wings, which 'appeared crazy,' she said, laughing. Moreover, although one usually escapes to the 'virtual' world to run away from frustration in the 'real' life, sometimes online life can, ironically, also put one in a bad mood, which will affect how one behaves in the 'real' life, as can be seen in this quote from Liuli (21-year-old, unemployed):

> Whenever I have killed someone, I feel extremely good. But I was killed not long ago by someone else, I felt extremely angry. I started quarrelling with my boyfriend. I was very unpleasant to him.

This indicates that even the activities in the 'playland' of the Net can involve much 'real' and 'serious' emotional investment, which can, in turn, affect one's self in the 'real' life, thus rendering the distinction between the 'real' and 'virtual' self even more problematic.

'NO. THE OPPOSITE.': THE INTERNET AND STUDY

A striking theme that has emerged from my interviews with urban youth is that they tended to see the Internet as antithetical, hence largely irrelevant, to studying (or other forms of what they referred to as the 'proper businesses', *zhengshier*). This view was also much shared by the urban parents and teachers I talked to in the past years, which echoes the widespread concern about the danger and risk related to the Internet in China, as mentioned in Chapter 2 and as I shall discuss in further detail in Chapters 5 and 6. When asked whether going online helped them in their studies or learning, the predominant reply from the middle school students was 'No. The opposite'. When asked whether they used the Net for school-related purposes, the reply was, 'No. we cannot even manage the knowledge in the textbooks.' When probed with the question: 'But there may be much learning involved in online recreation. Moreover, one can employ the Net for educational purposes. How can you say it is the opposite of learning?' In response, most of the informants claimed that there was anyway very little learning in *their* online activities (at least in the sense that it did not help enhance their exam results). This is what I heard at a group interview in 2008 with five Grade-1 students at upper secondary school:

> *Liliang (17-year-old, boy):* For us high school students, it is mainly an entertainment. We seldom use it to look up things related to our study, only when the teacher asks us to.
>
> *Zhangwei (17-year-old, boy):* Indeed very little. Seldom does the teacher ask us to use the Internet.
>
> *Zhouhong (17-year-old, girl):* But it can be useful sometimes. I sometimes visit the 'study' web.
>
> *Liliang:* But one does not need to do that if one grasps the textbooks well.
>
> *Liuhui (17-year-old, girl):* That is, in fact, quite true. The knowledge for us to learn is already very much, very comprehensive (*quanmian*) in the textbooks. We do not need the Net.
>
> *Zhaoqin (17-year-old, boy)* That's true. We have very little use of the Net at the time being.
>
> *Wangze (17-year-old):* The problem is even if one can learn from and with the Net, one is not tested in these things.

Such remarks point to the constraints incurred by China's exam-oriented educational system mentioned in Chapter 3. In such a system, one's academic performance is mainly or entirely assessed through paper-and-pen-based exams based on textbooks.

This wisdom was again widely shared by the high school students. Therefore, it was commonly understood that onc is supposcd to go online less and less, ideally totally avoiding the Internet, as one goes higher in the grade at senior high school, at the end of which the National College Entrance Examination, with its fatal effect on one's future life chance, will take place. Preoccupied with the preparation for this highly competitive exam, the high school students tended to live in the future tense concerning their relationship with the Internet, much in line with the pragmatic tendency among Chinese youth discussed in Chapter 3 to suspend many wishes and desires for the sake of the preparation of the Exam. Admittedly, they did recognize the Internet's potential for a broader range of purposes well beyond entertainment, as is reflected in remarks such as 'The Internet is everything. There is everything on it. One can get everything one wants there. One can do everything on it', 'The Net is a 'know-all' (*wanshi tong*), and 'One can go to it for help with everything'. However, the high school students meanwhile indicated that they could only wait until after the exam, when they would be 'more relaxed' (*geng qingsong*), in one informant's words, to integrate the Internet more closely into their lives. And it was emphasized that this 'integration-to-be' would be necessary in order to keep up with the 'knowledge/information society'. In the following chapter, I will describe the high school students' 'imagined Internet future' in further detail.

Nevertheless, this dream about the future may prove to be misleading, given that even when one has entered college, one may still use the Net mainly for recreation, as is reflected in the case of the three college students included in this chapter and the university students I interviewed for the topic of young people's online political participation (see Chapter 8). Reflecting the general online culture discussed in Chapter 2, these university students also tended to treat the Internet mainly as a recreational device. By the same token, those who were preparing for the Postgraduate Exam (*kaoyan*) deemed it necessary to keep a proper distance from the distracting Net. Zhengyuan (23-year-old, college student) told me that she rarely went online because she was bent on entering for the Postgraduate Exam (*kaoyan*), implying that once she went online, it would most likely be about things that were irrelevant or even contradictory to this goal.

Of course, as not all manage to exert sufficient self-control required for the rational choice, how far people succeed in such 'abstinence' varies from individual to individual. As some of my informants told me, the greater the pressure to work hard, as it is the usual case in the last year of high school, the more one may feel the need to relax by escaping into the 'virtual world'. The more one spends time online (given the predominant entertainment-oriented uses), the worse one's school performance becomes, so goes the vicious circle.

The widely shared perception of the Internet as the antithesis to study/work also explains why parents are extremely nervous about their children going online. Schutz (Schutz and Luckmann, 1973) argues that the everyday lifeworld is intersubjectively constructed by the subject and his or her fellowmen. This is not least true of technology adoption, ownership and access/availability within the family (Livingstone, 2002). As revealed by both my interviews with urban youth and my interaction with Chinese families, parents have been playing an important role in shaping the availability of the Net at home. Besides economic considerations, PC/Internet provision within the family depends on the extent to which parents view the Net as helpful for their children's overall development. As mentioned in Chapter 2, like parents in Western societies (Facer et al., 2001), Chinese urban parents, who typically cherish high expectations of their only-children, tend to see the computer/Internet as of great importance for their children's future, in line with the official discourse of the information society and knowledge economy. However, due to the fact that the current exam-oriented education system tends to leave little room for exploration beyond textbooks, they widely perceive the Net as antithetical to academic performance, especially given the predominantly recreational uses observed among Chinese children and young people. Thus, in the Chinese case, the Internet is simultaneously seen as highly relevant and absolutely irrelevant.

As a result, it is not unusual that parents are hesitant in having the Internet installed at home. In many families, although after much negotiation, the Internet has been installed, it is commonly understood by the family members that it is a collective gadget and part of the home interiors, like the family TV. It is thus typical that it is placed in the sitting room, rather than a private bedroom. Moreover, the home Internet is constantly under parental watch. Compared with their Western counterparts, for example, Norwegian youth (Liu, 2010), who typically have PC/Internet in their own bedrooms, the Chinese youth's bedrooms (and homes) then constitute a relatively poor 'media environment' (Livingstone, d'Haenens and Hasebrink, 2002). However, this difference cannot be purely explained from an economic perspective. Given the 'child-centered' culture and the purchasing power of the only-child family, most Chinese parents would have provided their only-child with a personal PC/Internet if they were convinced about its relevance to the child's future success. But, in the Chinese context, the youngster's bed room is an extension of school given the large amount of school work a student is supposed to do after school (and the small amount of free time). Home and one's own room are much less construed as leisure spaces than in some Western contexts, such as in Norway (Liu, 2010), where young people's 'bed-room culture' allows for a diversification of tastes and habits at home and represents an expression of individualized lifestyles on the part of young people (Livingstone, 2002). With the collective central goal of getting the child into a university, preferably a good one, the urban Chinese family life is much more organized around the child's study (more exactly, school-related work). Thus, the home Internet serves as a mere symbol of modern material wealth and a symbolic link to the information society in the newly emerging middle-class family. In other words, it is still beyond their zone of primary relevance.

This shows that in the Chinese case the Internet's relevance as an educational device, which is imposed by the discourse of the information society, is largely rejected as young people are 'forced' to choose between using it as *either* an educational *or* a leisure device, rather than both. And yet, 'choosing' to use it for the latter, as Chinese young people tend to do, is no less worrying for parents and teachers (and sometimes young people themselves), who are all highly concerned about young people's academic performance. Such a complexity needs to be understood within the exam-oriented educational system that frames young people's everyday lifeworlds.

CONCLUSION

How are Chinese urban youth actually taking up the new medium and making it meaningful for themselves? In this chapter, I have explored this question by analyzing my informants' narratives from a Schutzian perspective that allows for a 'situated' understanding of individuals' choice and meaning making in relation to the Internet. It serves as an illustration of how concrete individuals are actually constructing the Internet as an element of their everyday lifeworlds embedded within both the larger context of today's China and within their more immediate local context, reflecting both the general characteristics of urban youth and their local embeddedness of being young in today's China.

The major themes that have emerged from their narratives reflect their perceptions and lived experiences of the Internet in some major domains of their lives: recreation, the self, sociality and study. These themes together offer us a wholesome picture of how the young people perceived and experienced the Internet. It shows that they assigned the Internet a strikingly high degree of relevance to their recreational and expressive needs, but much less relevance to other purposes. Furthermore, their particular 'recreational' relationship with the Net has both shaped and been shaped by the relevance and meaning of the Net they assigned to the other major domains of their lives. This is particularly true of study (and by extension, learning, work and other 'proper' businesses of life), to which going online was viewed largely as an antithesis. Thus, emerging from their 'I-and-the-Internet' narratives is a predominantly expressive relationship with the new technology. The metaphors they used for the Internet such as the 'virtual world' and 'player' invoke a user-Internet relationship in which the Net is seen as a place and a way of being rather than a tool (Markham, 1998).

Such an orientation toward the Internet may not be surprising given that, as shown in Chapter 3, Chinese urban youth share an eagerness for individual self-expression. Moreover, my informants' particular relationship with the Net is not only indicative of the general tendency among children and youth to indulge themselves in playing in various forms, but also the kind of material welfare Chinese urban youth enjoy, as indicated in Chapter 3. The fact that they had access to this most advanced form of new media and that they could afford a regular visit to the Net café, or even 'steeped themselves in the Net Café' many hours daily, as some of them did—mainly for recreation—are indeed manifestations of their social-economic status in Chinese society. However, a closer look shows that behind such a seemingly hedonistic approach to the Net is the persistent theme of pressure, frustration and boredom, all of which is directly related to their social-biographical situations as Chinese urban youth, whose lives are further situated in the local context of a relatively less developed city. Thus, characteristics such as 'being young, Chinese, in a relatively small city, only-children'—factors that remind one of the discussion of today's Chinese youth in Chapter 3, were perceived as the main structural factors that had led to a strong need to escape to the 'other world' of cyberspace. In making account of their predominantly expressive relationship with the Net and assigning the relevance and meaning of the Internet to their individual lives, my informants drew different aspects of these situational characteristics into the foreground.

The shard perception among my informants that the Internet is counterproductive to academic performance indicates that in the Chinese case, the imposed relevance of the Internet as an informational and educational device seems to have largely been rejected. Or, it has become an 'imposed irrelevance' due to the very educational system itself, which, exam-oriented and textbook-based, has shaped the relevance and irrelevance of various sorts of information and knowledge. Under such circumstance, although it may be the case that by 'playing' on the Net, young people are gaining important knowledge and skills for participating in our contemporary and future society (Gee, 2000, 2003; Gauntlett, 2010/forthcoming; Hull, 2003; Lankshear and Knobel, 2006; Leu et al., 2004), these are seen as largely useless as they do not contribute to one's 'success' in the Chinese context. In other words, such a paradox can only be understood by locating young people's (and their parents') perceptions and experiences of the Internet within their co-constructed everyday lifeworlds embedded within the wider socio-cultural context of today's China, where the aforementioned exam-oriented educational system works as an important factor shaping their relevance system.

The major themes that have emerged from this chapter will re-emerge time and again in later chapters of this book. So far, we have seen how the Chinese urban youth in this study took up the Internet and made it fit into their relevance systems in ways that reflected their socio-biographical situations. However, their Internet-related narratives did not stop there, that is, they did not merely describe their particular relationship with the Net and offer explanations for that, but also went a step further to make moral evaluations of their own and other people's relationships with the Net. This constitutes another dimension of their Internet-related perceptions and experiences which I will take up in the following chapter, where the two contrasting camps of users mentioned earlier on in this chapter will be brought together.

5 Internet Anxiety, the Norm of the 'Good' Netizen and the Construction of the 'Proper' Wired Self

This chapter continues to explore young people's experiences and meaning making in relation to the Internet, but from a different angle than in the last chapter. Here I focus on how the participants' Internet-related subjectivities and self-understanding reflect the Chinese discourse of modernity, with the Internet as a central symbol, and related to this, the ideal subjecthood embodied in the double subjectification regime discussed in Chapter 1. I start by offering an account of the widespread anxiety about the Net in Chinese society.

INTERNET ANXIETY IN CHINA

The Internet's attraction of children and young people has been a major concern in China over the past years. Both academic and lay discourses view Internet addiction as a serious and widespread problem. The Net café has also become associated with 'negative influence' on the young. The cultural semiotics of the Internet are characterized with expressions such as '*wangyin*' (addiction to the Internet), '*chenjin*' (immersed), '*chenni*' (submerged), and '*chenmi*' (infatuated, addicted, or even in a coma) (Golub and Lingley, 2008). In an extreme version, the moral panic aroused by the Internet has led to a popular view of the Net as akin to addictive drugs, in particular opium, as one can easily check out by a search for 'the Net+opium' on the Chinese web. This phenomenon has also been observed by some other researchers (Bucher, 2004; Golub and Lingley, 2008; Tsui, 2005).

Because of its historical connotation, associating opium with the Internet triggers an avalanche of concern and sentiments (Tsui, 2005). That is, '[t]he real threat is not concerned merely with the bodies and minds of those addicted but also their family and society's moral order' (Golub and Lingley, 2008: 6). This is the argument made by Zhang Chunliang, China's well-known anti-Internet crusader:

> Without exaggeration, a war to win over the young generation has started. This is a war without smoke from weapons. Shocking realities

> of blood and tears make it clear that this is a war fought over China's future and Chinese nation's fate. Therefore, it can be said that if not controlled well, online games will most likely become the 21st century's 'electronic opium' and 'spiritual opium'.[1]

Such worries about the Internet's 'harmful' effect echo the general concern about the so-called 'moral slide' so resonant in the Chinese discourse about modernity and risk, and related to this, 'socialism with Chinese characteristics' and the 'two civilizations'. As indicated earlier in this book, the Chinese leadership views modernity as signifying both hope and danger. Then, it is not surprising that the Internet, as an ultimate symbol of modernity, has always been viewed with the same ambivalence by the government. Reflecting its dual attitude toward the Internet mentioned in Chapter 2, the government sees a whole generation of tech-savvy youth as central to China's long-term interests in the global economy. However, it deems it essential to guard the younger generation against the 'harmful influence' of the Internet. The Internet, with its open and networked nature, makes transmission of information and communication much faster, broader and more multi-sourced than traditional media. As such, it makes it harder for the government to 'protect' the Chinese population from the so-called 'Western cultural influences' in both political (such as values related to democracy and human right) and moral (such as liberalism, individualism, materialism, hedonism and pornography) terms, both being the targets of the 'socialist spiritual civilization'. The worry is that the Internet reinforces the diversification of value systems, liberalization, individualization and fragmentation of Chinese society—contrary to the socialist-collectivist values with which the Party state is still trying hard to inculcate its population. Therefore, guarding the Chinese people, especially the young, against information and social spaces in cyberspace that can 'contaminate' their minds is a central goal of the moral education designed for 'socialist spiritual civilization', of which the 'harmonious society' discourse is an extension. Moreover, for the government, young people's tendency to indulge themselves in the virtual world, using much of their productive time, represents a form of the hedonism and loss of control indicative of the 'moral slide' accompanying China's reform and opening up. In the social engineers' terms, the moral consequences of the Internet are detrimental to both the 'material civilization' and the 'spiritual civilization', both of which are crucial to China's modernization. Thus, concern about the Internet's 'bad effect' on the social-political-moral order underlies the state's enormous effort to control the Internet.

The general public seems to be equally ambivalent about the Internet. Particularly, parents find themselves in a further dilemma besides sharing the general concern about the Net's 'harmful' effects on their children's overall development (Liu, 2009, 2010). On the one hand, they see the computer/Internet as an investment in their only-child's education with implications for the family's social mobility, in line with the discourse of 'information society'. On the other, since the current exam system leaves little room for exploration beyond textbooks, Chinese parents widely perceive their youngsters' time on the Net as antithetical to learning and academic achievement, especially given the predominantly recreational uses among Chinese children and young people. Moreover, parents' concern about the Internet's distraction of the young from school work is combined with worry about losing authority over their children. According to popular discussions, Chinese youth are becoming more and more self-centered, pleasure seeking, lacking in self-control, hedonistic, irresponsible and disrespectful to their elders. There seems to be a shared understanding among Chinese parents that the Internet, particularly when accessed at the Internet café, is part and parcel of such deterioration of morality among the younger generation. Such parental worry coincides with the aforementioned concern among Chinese leaders about the general 'moral slide' of society in light of China's modernization.

The widespread Internet anxiety is reinforced by media coverage about young people and Internet addiction, typically with an emphasis on its destructive effect on the mental and physical health of the user and hence an urgent call for societal attention. For example, according to the third *China Youth Internet Addiction Report* sponsored by the China Youth Internet Association and published in early 2010, 24 million, namely, 14.1% of China's young netizens, are addicted to the Net. The report is based on a survey of the 4 national municipalities and 26 provincial capitals—all the major cities in the country. According to the report, the number of those suffering from Internet addiction has nearly doubled since 2005 when the first such survey was conducted. Online game addiction (47.9%) was the major form of addiction, followed by the so-called 'cyber-relationship or networking addiction' (chat and/or making friends) (13.2%). It further shows that young people aged 18–23 years constitute the highest proportion of the addicted youth, followed by people 24–29 years of age. Compared with 2005, the proportion of addicts aged 13 to 17 years has declined and that of the 18–23 age group has increased. The report claims that this is thanks to the efforts by the state departments in protecting the minors online. Besides, the report also makes an association between 'moral slide' and 'addiction', citing the statistics that 90% of the non-addicted youth were against 'pornographic transactions', whereas among the addicted youth, as many as about 95% deemed it acceptable and that concerning online one-night stand, 76% of the non-addicted youth thought it unacceptable, whereas about 90% of the addicted youth believed it to be acceptable.[2]

The widely shared concern about the pitfalls of Internet use is also reflected in the large scale surveys led by Guo Liang at the Chinese

Academy of Social Sciences. For example, the report published in 2007 (Guo, 2007) based on a survey in seven cities with a sample of 2,001 urban residents between the ages of 15 and 59 (1,315 Internet users and 686 non-users) shows that an overwhelming 93% of the participants considered much of Internet content to be unsuitable for children. About 61% thought Internet users could easily become addicted to the Internet, and the same number thought users could easily be affected by online pornography. Over 43% said the Internet could lure users into making the wrong kind of friends, and another 42% said Internet use easily presented risks to personal or private information. Moreover, these negative impressions were significantly stronger among non-users: 72% were concerned about pornography, 81% about Internet addiction, 66% about making the wrong kind of friends, 55% about risks of exposing personal information.

In addition to research reports, the Chinese media have been replete with stories of cyber victims in recent years, although reliable statistics are hard to come by. The bizarre and pathological behaviors related to the Internet that have featured prominently in the Chinese media since 2002 (Golub and Lingley, 2008) range from suicide to running away from home, stealing and robbing and homicide (even one's own parents or kin). Just to name some examples: A 13-year-old boy leaped to his death from his family's 24th-floor apartment in Tianjin after playing online games for 36 hours straight (according to his diary, he had 'flown to the sea' to join his friends in the virtual world); a man stabbed another in the chest after the victim sold a hard-earned and coveted virtual sword won in cyberspace to another player; a boy murdered his uncle to steal money to play video games; two children exhausted by gaming marathons fell asleep on railroad tracks and were killed; a man murdered another over the theft of a virtual sword in Mir II; four teenagers were seriously injured in a knife fight between 19 gamers at an Internet café over treasure dropped by a monster in a massively multiplayer online game; a 17-year-old killed himself after meeting his online sweetheart 'Flying Skirt' in real life, only to find she was a homely woman in her 30s; a young man from the Southern city Wuxi killed his father and severely injured his mother after his father cut off electricity to stop the son from playing online games. Other youths reportedly have dropped out of school, or have been jailed for petty crimes carried out to support the gaming habit (see Golub and Lingley, 2008; Global Technology Forum, November 9, 2005).

In recent years, alarmed by such media reports, the government has engaged in a nationwide campaign to stamp out what the Communist Youth League calls 'a grave social problem' that threatens the nation. An anti-Internet addiction system has been in place since 2007, requiring domestic online game operators to install a real name checking system to prevent minors from becoming totally immersed in the virtual world of online games (Lo, 2009). This system basically tracks the online use by minors, using a credit system which responds negatively when any minor is playing online games for more than 3 hours a day (Block, 2008). The government claims that the system is effective in curbing online gaming addiction, citing the statistics released by the Youth Center Services in China which shows that the number of online gamers under the age of 18 dropped from 18% to 15% from the year 2007 to 2008.[3] However, it seems that such regulations are easily circumvented given the relentless efforts of Internet service providers to create a market of mass consumption and the great need widely perceived among Chinese urban youth to escape into the virtual world from the great pressure, boredom and tight control in their everyday lives.

In November 2008, the Chinese government officially defined Internet addiction as a psychiatric disease and it has been sponsoring some clinics, or rehabilitation centers, to treat addicts (*The Guardian*, November 11, 2008). Many parents have sent their addicted teenagers to such clinics, at which some of the teenagers are subjected to electroshock therapy so they will be 'reborn' free of the obsession (*Timesonline*, June 7, 2009). According to the *Washington Post* (February 21, 2007), the clinic in Daxing, a suburb of Beijing, the capital, is the oldest and largest, with 60 patients on a normal day and as many as 280 during peak periods. Few of the patients, who range in age from 12 to 24, are there willingly. Most have been forced to come by their parents, who are paying upward of $1,300 a month—about 10 times the average salary in China—for the treatment. Led by Tao Ran, a military researcher who built his career by treating heroin addicts, the clinic uses a tough-love approach that includes counseling, military discipline, drugs, hypnosis and mild electric shocks.[4]

'Softer' methods are also employed to educate young people about the harm of Internet addiction. For instance, '*Shan Dian Mao De Gu Shi*' (The Story of Shan Dian Mao), the first Internet addiction themed television show in China, put on in 2005, was meant to reveal to teenage viewers the horrors of using the Internet too much and to teach youth ways to avoid addiction. This show is based on the life stories of a number of Internet-addicted 'newer new humanities', as the present-day Chinese youth are called (see Chapter 3). Like the Chinese media in general, it is rife with stories about youngsters who fall ill or die in Asia because they spend too much time playing games or chatting online. The 38-episode sitcom was a joint effort by the Care for the Next Generation, China Youth League's Online Movie and TV Center, and the China Youth Internet Association. According to *China Youth Weekly*,[5] the show attracted wide attention and marked the start of a nationwide 'resist Net addiction' campaign. In the wake of this show, a large-scale seminar with the thematic slogan 'Fight against Net addiction, save the child, strengthen our base and solidify our roots, and harmonize the

Chinese nation' was held at the Great Hall of the People.[6] Here again, we see a clear link between Internet anxiety and the larger social-political concerns in present-day China.

In short, the widespread concern about young people and the Internet in China directly confronts how children and young people are actually using the Internet in contention/tension with parental and societal expectations. The anti-Internet discussion in China typically combines a scientific discourse (e.g., negative mental and physical effect), a moral discourse (erosion of ethics and morality) and nationalistic discourse (concern about the quality of the future builders of the Chinese nation). All of these emphasize and call for a balanced, self-disciplined approach to the Internet, suggesting a certain norm about how one should relate to the Internet. Such discussion is reminiscent of the 'two civilizations' and moral education, with an emphasis on individual self-control in line with the double subjectification regime. As a norm, it attains discursive power which the user has to come to terms with somehow.

NEGOTIATING THE 'PROPER' WIRED SELF

In the rest of this chapter I shall explore how the earlier-mentioned norm is being played out in Chinese urban youth's negotiation of a 'proper' wired self. I draw on the narratives of all the 70 participants described in the preceding chapter. Following Grant (2005), I use 'proper' to signify the unnaturalness, that is, the negotiative nature of my participants' subjectivities, related to the moral order of society. I use 'wired self' to indicate the participants' self-understanding as members of the so-called 'Internet generation', also referred to as the 'wired generation'.

Drawing on the notion of 'interpretative repertoire' proposed by Potter and Wetherell (1987, 1995), I explore two questions: What kind of interpretive repertoires can be identified in the participants' accounts about their relationships with the Internet? What kinds of subject positions do the speakers take up for themselves (and assign for others) by drawing on these interpretative repertoires?

Potter and Wetherell (1995: 89) define interpretative repertoires as 'broadly discernible clusters of terms, descriptions and figures of speech often assembled around metaphors or vivid images' which are 'available resources for making evaluations, constructing factual versions and performing particular actions.' Potter and Wetherell (1987) further emphasize that descriptions and evaluations are inseparably bound together with larger cultural models of meaning making. In producing accounts, people are often displaying knowledge of the ideal personhood expected of them. They talk about themselves in terms of the social order in which their lives are embedded, thus making their behavior accountable in a moral world (Shotter, 1984). As such, identifying the particular interpretative repertoires my informants drew on in their narratives serves as a useful way to understand how the ways in which they make accounts of their own and other people's Internet-related experiences are related to the wider moral system of reform-era China.

Furthermore, interpretative repertoires provide subject positions in discursive practice (Bamberg, 2004; Davies and Harre, 1990). A subject position offers a specific identity to which the speaker has to relate herself and hence a vantage point from which she sees the world (Davies and Harre, 1990). However, it is important to adopt a more actor-oriented notion of the subject actively positioning herself, rather than a deterministic notion of the subject 'being positioned' (Bamberg, 2004; Wetherell, 1998). In this way, we would see a stronger social agency, a greater control of the discourses one deploys and a greater ability to reshape these for one's own purposes.

I have identified three interrelated interpretative repertoires. Each of these is characterized with a dual pattern that resonates with the tension between the 'approved' ways of relating to the Internet and those that are often viewed with doubt and anxiety.

The 'Meaningful versus Meaningless' Repertoire

In Chapter 4 I show that the majority of the young people interviewed reported a predominantly recreational orientation toward the Net. Their narratives invoke a user-Internet relationship in which the Net is seen as a place and a way of being rather than a tool (Markham, 1998). Metaphors such as the 'virtual world' and 'player' were frequently employed by the participants to convey such a perception and experience of the Internet. This forms a sharp contrast with a few (6 out of 12) of the older informants (aged 25–28), who reported using the Internet mainly for purposes related to work and study—uses that invoke the notion of the Internet as a tool. From a Schutzian perspective, such a difference in itself is not surprising given the different configurations of various factors that shape one's social-biographical situation, such as age, gender, social-economic status (e.g., being employed or unemployed, family background, student and non-student status) and marital status. Different social-biographical situations mean different relevance systems with differing hierarchies of plans. Indeed, as we shall see throughout this chapter, in their accounts of their relationship with the Internet, these 'instrumental' participants did explain their relationship with the Internet by linking it to their everyday life actualities, just as the 'expressive' users did, as I have shown in Chapter 4. However, what strikes me is that the two types of users deliberately distinguished themselves from each other, camping themselves into two contrasting types of users. Those who used the Internet mainly for entertainment defined themselves as 'players' (*wanjia*), as mentioned in Chapter 4, or 'recreational users'. In contrast, the few in the other camp saw themselves as the 'pragmatic' or 'instrumental' users. In doing so, the participants tended to draw on a dual interpretative repertoire that poses various uses of the Internet as either 'meaningful' or 'meaningless' (alternatively, 'useful' or 'useless'). Moreover, the two camps largely shared the understanding that the 'instrumental uses' are 'useful or meaningful' whereas the 'recreational' uses are the opposite. The dual nature of this 'meaningful vs. meaningless' repertoire, like the 'good vs. bad' repertoire which I will discuss in the following section, creates a discursive space in which although one may employ the Internet mainly for fun, the opposite set of expectations, which is often perceived as the correct use, is present, and vice versa.

The ways in which the two types of users distinguished themselves from each other echo the dichotomic distinction made in earlier theorizing of user-computer relationships (e.g., Turkle, 1984; Aune, 1996) between 'instrumental' and 'expressive' uses. Such a distinction can often be questioned given that in reality the various uses are often mixed rather than dichotomized. Nevertheless, it is exactly such a dichotomic thinking among the young people that is worth exploring.

Just as the few 'instrumental' users defined their uses as 'useful and meaningful', the 'recreational users' saw their own engagement with the new medium as at best 'counter-productive' to school work in the case of the students, or as a way of killing time or being idle in the case of vocational school students and the non-student young adults. When asked how important the Net is for them, Liuxiang (16-year-old girl) indicated, 'It is very useful in some ways such as entertainment, but it can also be seen as useless because it delays (*danwu*) study.' As I have shown in Chapter 4, the 'recreational users' were quick to add that using the Internet mainly for recreation despite of its 'uselessness' was specific to their ages and life situations characterized with great pressure, lack of space for individual expression and boredom. Or as Xietian, a high school boy emphasized at a group interview, 'For people at our age, it is just entertainment. We seldom do useful things with it', implying that it may be different for other people and that he and his friends knew well about the other possibilities with the Net.

As already indicated, Xietian's notion that online recreation is useless was widely shared among the informants. Thus, although via online activities they were picking up basic social and technological skills necessary to fully participate in contemporary and future society (Gee, 2000, 2003; Hull, 2003; Lankshear and Knobel, 2006; Leu et al., 2004; Sefton-Green, 2004), the recreational users were largely ambivalent about the value of these skills. They indicated that being able to use the computer/Internet was something that distinguishes their generation from their parents, which made them feel privileged, because 'this after all signifies social progress', as Linke (20-year-old, man, vocational school student) argued. However, as this did not contribute to one's academic achievement (measured solely by paper-pen exams based on the textbooks) and other more obvious forms of productivity (such as moneymaking) in a highly market-driven consumerist society, the young people found it hard to attach much relevance to their computer/Internet skills. Furthermore, much in line with Gee's (2003) argument about online gaming and learning, besides the basic computer/Internet skills, some of them recognized that online gaming fosters one's cooperative spirit and social skills, as well as quickening the mind, as one has to make strategic moves all the time in the games. Besides, many of them had learned to use complex multimodal texts to decorate their online personal spaces, which constitutes an important form of 'new literacies' (Gee, 2003), or, in Gauntlett's (2010/forthcoming) terms, an important way of 'making' which serves as 'connecting', with great implications for our engagement with the world. However, echoing the societal concern about youth's 'misuse' of the Net, the recreational users saw these 'new literacies' as largely not conducive to the goals of the 'good life' in the future, as they served no other function than for fun.

As we can now see, drawing on the dual interpretative repertoire of 'useful/meaningful vs. useless/meaningless', the recreational users positioned themselves as the technologically competent, but pleasure-seeking youth who employ the Internet almost purely for fun although they know well about the other possibilities with the Net. Such a subject position allows them to talk, often in a somehow self-defensive way, from the vantage point of a young person at certain life stage and with their cultural baggage as only-children living in the specific socio-cultural context of today's China. Moreover, the dual interpretative repertoire allowed them to position people who use the Net for instrumental ends as 'people who are not in the same situation as we are', and who, therefore, have a different orientation toward the Net.

As in the case of the recreational users, the way the instrumental users described their relationship with and perceptions of the Internet shows that the dual interpretative repertoire of 'meaningful versus meaningless' has a strong impact upon their subjectivities. Just as the recreational users deliberately stressed that using the Net for entertainment is specific to *people at their age,* the instrumental users intentionally distinguished themselves from those who use it mainly for recreational purposes. That is, they seldom talked about themselves without referring to the vast numbers of young people who indulge themselves in online recreational activities and worse still, even 'bathe themselves in the Net Café all day long' just for 'meaningless things', in their words. Like the recreational users, they largely saw online entertainment as a waste of time and an indication of idleness. They claimed that 'people like us', in their words, use the Internet mainly at home and work places and almost purely for 'useful' things, such as things related to work and study, looking up information, keeping in touch with far-away relatives and friends, making photo albums, e-shopping and e-business.

In defining the Internet, the instrumental users saw it as an enabling tool. For example, Wangze (25-year-old man, college librarian) remarked, 'The Net is an extender of your hands and eyes. So, it is something that makes some things in life easier to do and cheaper.' For Dulei (27-year-old woman, kindergarten teacher), 'The Net is a *gigantic* library.' A 28-year-old secondary school teacher mentioned that she often found useful material for her teaching work. The instrumental users' perceptions of the Internet are in line with the emphasis on utility and efficiency that has informed the public rationales and cultural justifications for both public and private investments in access (Sandvig, 2006).

By drawing on the same dual interpretative repertoire, the instrumental users positioned themselves as the competent, rational, pragmatic, mature and responsible user. Moreover, by posing the recreational users as their 'opposite' or the 'other', they positioned the latter as the irrational, immature and irresponsible users, a position similar to that sometimes taken up by the latter themselves. Whereas the subject position taken up by the recreational users seems to be more complicated, as we shall see later on, that taken up by the instrumental users is consistent throughout their narratives, and so is the one they assigned to the recreational users.

The 'Good versus Bad' Repertoire

As one can already discern, my informants tended to make moral assessments about their own, and other people's, relationships with the Internet. I noticed this tendency in almost every interview, where as soon as I mentioned that this project is about young people and the Internet, I got the comment: 'The Net is both good and bad.' Or as a young man put it, 'The Net is a double-edged sword.' In doing so, they tended to frame their narratives around another dual interpretative repertoire: the 'good vs. bad', which alternates with similar ones such as 'proper vs. improper' and 'right vs. wrong'. Realizing that this might be due to the possibility that they viewed me as an authority figure seeking self-reflection from them (although I did feel that our conversations were going on in an informal and relaxed atmosphere), I would deliberately explain that my main focus was not on value judgement of the Internet. Rather, I was more interested in how young people are actually taking up the Net in their lives. However, I still noticed a strong interest in commenting on the 'good' and 'bad' of the Internet among the informants. Their persistence in the topic and the fact that they assumed that everyone investigating young people and the Internet is focusing on the 'harmful' effects reflects the far-reaching Internet anxiety in Chinese society.

This interpretative repertoire seems especially resonant when the high school students talked about going online and study. As indicated in Chapter 4, a shared understanding among the participants was that a good student does not go online (much). Or as Zhangyi (17-year-old, student), who often visited the Net café, put it when representing her friends at the same interview: 'We are not good students. Good students seldom go online', a remark also made by some of the young adults mentioned in Chapter 4.

Accordingly, I was told time and again by both those who defined themselves as good students and those who defined themselves as otherwise that although nowadays one seldom finds an urban youngster who does not go online at all, 'good students' use the Internet in very self-controlled ways, just as they spend very limited time playing in other forms. This argument got further support from some of the older informants, who cited themselves and others to demonstrate a law that a 'good' student will become a 'bad' one by going online (too much). For example, Zhangfeng (a 19-year-old man who had just graduated from high school and was ready to go to college at the time of my fieldwork) told me:

> I was one of the best students in the beginning. But my school results fell down dramatically because I played (online) very much in high school. Therefore, I failed the college entrance exam last year and had to repeat a year to retake the exam this year.

In a similar manner, Zhaoli (21-year-old, vocational school student) told me that she was originally a very good student. But because she played online so much during her junior high school years, she failed even to enter senior high school and so ended up at a vocational school, which does not promise a bright future for her. Some emphasised that such stories are numerous in society. This reminds one of the many cases in Chinese society that are constantly reaching one's ears and eyes via the media and from people around oneself, all pointing to the Net as a serious distraction of young people from their school work.

Thus, those students who went to the Net café whenever they were 'free' seldom did so without a sense of guilt or uneasiness, as 'this is after all bad for a student', as some of them indicated. However, when they were playing online just in order to relax from school and parental pressure, as they told me—compared with those who 'steep themselves all day long in the Net café', in their words—they were not that 'bad'. Or as a male student put it, 'We are just ordinary netizens. We are not as bad as those who submerge themselves all day long in the Net café'. Likewise, they viewed their classmates who were disciplining themselves for the purpose of exams and who therefore rarely went online with a mixed feeling of contempt and admiration, as is clear in the following interview excerpt:

> Such people appear unsocial and somehow strange. I do not approve of that type myself. But on the other hand, I admit that they are doing the 'right' thing. After all, your goal is to enter the university (*kaoxue*). When you have entered the university, in fact, it is very fast to learn that thing (using the Net). (Tanyong, 16-year-old, boy)

By the same token, the few high school students who told me that their parents did not control them that much concerning the Internet, indicated that an important way to live up to their parents' expectations was to exert self-discipline in going online. Thus, the first step toward becoming a 'better student', hence a 'better child', is to go online less.

I discerned a much more relaxed assessment of their own relationship to the Internet among the 'heavy recreational users'. These consist mostly of the young adults who were vocational students and those who had left school and were waiting for employment at the time of my fieldwork. For these youth, an academic career is hardly relevant now. They told me they belong to the type that does not like studying. Therefore, they were spending much time at the Net café (5–10 hours daily), mainly playing online games. However, although they viewed this as just normal for people like them, they meanwhile reflected upon their orientation toward the Net along the dual repertoire of 'good vs. bad' just as the previously mentioned high school students and college students did. For example, Wulong (a 22-year-old man at a vocational school) said, representing the friends present at the same interview:

> We know we should not play so much. It delays the proper businesses in life. It can also be very bad for the eye and the health in general. I once vomited after having played overnight at the Net café. But it has become a habit for us to go online to play games (Laughing).

Liwei (25-year-old man, railway station employee) was even more explicit in this respect:

> The Net is very useful for looking up things. There is everything on the Net. But if one uses it just for playing, it is very bad. I feel my health is failing because of that. I get headaches after playing a few hours on end and lose my appetite. Moreover, it is not good for your spirit either. As the saying goes, *wanwu sangzhi* (one loses one's will by indulging oneself in playing with things). Therefore, my parents have been against my playing online games from the very beginning. But it is hard to pull myself out.

Thus, the dual repertoire of 'good vs. bad' allows the 'heavy recreational users' to reflect upon their own use of the Internet with a critical self-awareness. In doing so, they positioned themselves as the technologically competent, but irrational and weak-willed addict to the Net, whose behavior should be corrected. The same dual repertoire offered a discursive space within which the 'moderate recreational users' could position themselves as users who were 'not that bad' since they could exercise self-control, although they were not the most rational users compared with the 'good' students. As such, making an assessment of various uses of the Internet, the recreational users' critiques were directed toward themselves.

In contrast, the instrumental users employed the same dual interpretative repertoire to problematize the behavior of people who use the Net purely or mainly for recreation, while defining their own, although not always explicitly, as the proper way to relate to the Net. The explanations they offered of why today's youngsters use the Net the way they do sound quite different from the explanations the recreational users offered themselves. Some provided a more sympathetic view, arguing that this is because 'they are still young', not 'understanding things' yet and hence unable to make the 'right' choice about the Net. However, some others sounded more critical, blaming the latter for 'not attending to the proper businesses (*buwu zhengye*)' and attributing their indulgence in the online world to their lacking goals in life and a general lack of good moral education both at school and at home. One instrumental user made an explicit association between this phenomenon and the only-child, claiming, 'Today most families have only one child and the child is so spoiled that he or she is good-for-nothing, only concerned about pleasure.' Another insisted, referring to the 'weak-willed' addicts, 'The Internet is just a tool. You should not let the tool control you. Rather, you must control the tool.'

Employing the 'good vs. bad' dual repertoire, the instrumental users persisted in positioning themselves as the rational, mature and responsible users, whereas the (heavy) recreational users were positioned as the opposite.

The 'Past versus Present/Present versus Future' Repertoire

As mentioned earlier, my informants identified themselves as either 'recreational users' or 'instrumental users'. However, they did not see one's relation to the Internet as a fixed entity. Rather, it was viewed as a changing process that links the present to the past and future. Thus, their narratives revolved around a third dual interpretative repertoire: the 'past vs. present/present vs. future', which works in tandem with the two aforementioned dual repertories to inform their self-formation. Generally speaking, this interpretative repertoire suggests that there is a turning point between immaturity and maturity in one's relationship with the Net that one is supposed to reach. And echoing the general criteria for maturity in the Chinese culture, the extent to which one was seen as having achieved this was defined by the extent to which one was 'playing' on the Net as opposed to employing it for 'useful' things.

For the instrumental users, the turn to maturity had already taken place. But not everyone experienced this transformation in the same way. For some, the change had been dramatic. For instance, three friends (all men) at a group interview told me that they used to play online games very much, even to the point of addiction, whereas now they seldom did that because they were busy with the 'proper' things in life, be it to make money, work or manage a family of one's own. Moreover, some of them indicated that as adults they were now bored with online games. For example, Qiuliang (27-year-old man, public servant) said with emphasis, 'It is also tiring (to play online games). Very boring. When we have time, several brothers (as men friends often call each other in China) would rather sit together and chat, than going online (to play).' The result of such changes was employing the Net mainly for instrumental purposes and spending only limited time online. Looking to the future, a shared understanding was that they would use the Internet even less than currently once they get married and have a child, unless one's work relies on or is related to the Net, a point confirmed by the few who were married. Besides, they reported that they seldom, if ever, went to the Net café any more. They largely saw this change as the effect of coming of age and a sign of maturity.

Thus, these 'mature' users now talked in the tone of the veteran who had retired from the online world, where they used to play all kinds of games and get to the top level in each of them, as they claimed. Interestingly, however, by 'all kinds of games', they meant the popular games made in China or the Chinese versions of Korea-made games. They indicated that due to language (and cultural) barriers, they found it hard to play English games, such as World of Warcraft.

Not everyone, however, has gone through a dramatic transformation in their relationship with the Internet. This can be demonstrated by the examples of Dulei (27-year-old woman, kindergarten teacher) and Liyan (26-year-old woman, accountant), both of whom indicated that they had never played online games much or used the Net 'excessively' for other forms of entertainment, nor had they visited the Net café much. For this reason, they referred to themselves as 'good children' (*hao haizi*), evoking the dual repertoire of 'good vs. bad'. Although they spoke half-jokingly, this does echo the social norm of what constitutes the 'proper' relationship with the Internet. They reported that compared with before, they were now using the Net even more for work and other 'useful' things and less for entertainment. Seeing this state as indicative of maturity, they tended to view it as their future relationship with the Internet too.

Again, the instrumental users, either the new converts or original 'believers', who have always been 'relatively mature', positioned themselves as the competent, rational and responsible users, and the (heavy) recreational users as the opposite. However, constructing the 'maturity turn' as largely a 'coming of age' matter by employing the 'past vs. present/ present vs. future' dual repertoire, they offered, albeit not in clear terms, a more optimistic view about the latter, as the time will come when things change for the better.

The role time may play in one's changing relationship with the Net was also widely recognized by the recreational users. Those who were still 'playing online' very much, and hence frequently visiting the Internet café, indicated that things were changing because they were about to do the 'proper things' in life soon, be it to start business via self-employment or to find a job upon completion of an educational program. For example, Zhaoyang (22-year-old, unemployed), who had been playing online most of his time in the past years, reported planning to return to his hometown to do business together with his father. Some of the vocational students started to pay attention to information online that could help with their job hunting. Although they did not predict a future with nothing to do with online recreation, they predominantly indicated that they would surely be 'playing' less, or simply 'using' the Internet less as they equated the Internet with recreation, as they would have to work and have other responsibilities. Here is what I heard from Hanxue (22-year-old woman student at a vocational school):

> As a student, I can play so much. But when I am working, I will be playing less. Unless one's job has to do with the computer/Internet, one will be using it less and less. For example, in our subject, dental technician work, we will have very little to do with the computer. Besides, as one gets older, one also gets less curious about the Internet. For example, before we used to chat with strangers sometimes, now we do not find it interesting to do so any more. We have grown up. Not as interested as before.

The couple of recreational users in their late twenties showed a sense of emergency that it was high time for them to stop using the Internet like that given the bad effect on their health and spirit, as in the case of 27-year-old Hetian mentioned in Chapter 4.

In short, these recreational users viewed their changing relationship with the Net as a gradually maturing process, echoing some of the instrumental users' view in this respect. It is noteworthy that such a notion of one's changing relationship with the Internet conveys a rather narrow-minded understanding of the scope of Internet uses. It implies that as one matures, one should use the Internet less, because the Internet in these young people's minds largely represents a mere leisure device.

The predicted trajectory of the high school students seems to be somehow different from the earlier-mentioned groups. However, the belief in change was equally strong. Those who defined themselves as just so-so students and who were employing the Net regularly for relaxation from school indicated that they had nearly said good-bye to online games. And one reason was that they, like the earlier-mentioned older 'retired gamers', had had their days in the gaming world. Therefore, they talked in the same 'veteran tone' as the former. But more importantly, they explained, the reason why one had to limit oneself in online entertainment (or generally speaking going online at all) was that they have entered, or were just about to start, senior high school, which meant that they would have to focus on preparing for the College Entrance Exam, which would take place in a few years' time. As this exam has served as the most effective mechanism stratifying young people in China, with great implications for one's future life chances, they indicated that they would, from the following semester on all the way till they had taken that exam at the end of the upper high school, have to go online much less or stop going online totally. But they foresaw that when they had completed secondary school, they would be free in using the Net. Such a notion of 'future freedom' was even more resonant among the 'good' students who had all the time exerted self-control in going online. That is, when they had entered college, they would make up for the 'loss' regarding using the Internet. They foresaw that they would integrate the Internet into their everyday lives for various purposes, rather than mainly for recreation, much in line with the discourse of 'information society', with information seeking for productivity as a major purpose of going online. However, the Internet-related narratives of my college student informants, both those described so far and those to be analyzed in Chapter 8, seem to indicate that the high school students' prediction about their future uses of the Internet as higher education students might prove to be over idealistic. The greater 'freedom' enjoyed by Chinese college/university students seemed to be mostly exploited in the realm of online recreation (and communication), as I shall show in Chapter 8. The imposed relevance of the Internet as an educational/informational device tends to lose its power even at higher levels of education largely characterized by paper-pen and textbook-based exams, in similar ways as at the various levels of schooling.

Employing the dual repertoire of 'past vs. present/present vs. future', the recreational users, like the instrumental users, were able to construct their relationship with the Net as a changing process rather than a static entity. Such a repertoire allows the 'heavy recreational users' to position themselves simultaneously as the irrational, weak-willed addict, whose behavior needs to be rectified, and as one who is optimistic about the future change. Drawing on the same dual repertoire, the 'moderate recreational users' continually positioned themselves as the rational and responsible users who could exercise self-control. Such a position is characterized with the ability to put off pleasure seeking for the sake of one's life goals. As such, it reminds one of what seems to be typical of Chinese only-child teenagers: bent upon the preparation for the College Entrance Exam, they tend to live in the future tense, instead of the present, as discussed in Chapter 3.

CONCLUSION

The Internet is not merely a technological object, but it serves as part of a web of values, relationships, symbols and routines that make up social life. As such, it is unavoidably loaded with competing meanings, which is reflected by the different, and sometimes conflicting, notions about what constitutes 'proper' uses. Whereas the tension between the socially 'approved' uses and those often viewed with doubt and concern is found in many societies, it may be manifest in ways specific to the Net's local sociocultural contexts.

In this chapter, I have explored how the tension between competing discourses about 'proper' uses in China is being played out in a number of Chinese youth's narratives in relation to the Net. Adopting a discourse analysis approach, I have shown that negotiating the 'proper' wired self, the young people actively drew on three interrelated dual interpretative repertoires. These serve as channels through which the informants display knowledge of the ideal ways of acting in their local society (Potter and Wetherell, 1987) and reflect upon their own relationships with the Net along certain lines. The duality inherent in these interpretative repertoires thus allows the informants to position themselves as either the rational, responsible and mature user or the opposite and people who use the Net differently from themselves as the 'other'. Moreover, depending upon the extent to which one's relationship with the Internet is socially 'acceptable', some of the young people had to negotiate more than others in trying to come to terms with the norm about a 'proper' relationship with the Net. Whereas the 'instrumental users' found a ready fit between their uses of the Net and the social norm, the 'recreational users' found it hard to reconcile with the norm and tried to justify their relationship with the Net by locating their narratives in their social-biographical situations. This difference is further reflected in the metaphors they used to describe the Internet, such as 'library' and 'extender' versus the 'virtual world'. Metaphors, terms and stories used to describe the Internet are value laden (Markham, 1998; Stefik, 1996). Some uses are valued and others devalued, thus perpetuating the dichotomic thinking about the Internet (Sandvig, 2006).

These dual interpretative repertoires revolve around an Internet anxiety which would be familiar to anyone who follows Internet addiction debates in some other countries such as the U.S. and South Korea (Golub and Lingley, 2008; Block, 2008). Furthermore, they reflect the tension between Chinese youth's predominantly recreational orientation toward the Net and the social expectations of the new medium for enhancing productivity. This is reminiscent of the tendency found in some Western societies, where children and young people are far more fascinated by the Internet as a communication and entertainment medium than as a learning and information resource (Buckingham, 2003; Burn and Willett, 2004; Livingstone, 2003; Livingstone and Bober, 2004; Sandvig, 2006; Sefton-Green, 2004; Seiter, 2005). Despite growing recognition of the 'new literacies' emerging from children and young people's interaction with the Internet (Gee, 2000, 2003; Hull, 2003; Lankshear and Knobel, 2006; Leu et al., 2004; Street, 2003), adult concern about the Internet's 'counter-learning' effect also persists in the West (Buckingham, 2003; Seiter, 2005; Sandvig, 2006).

Nevertheless, it would be misleading to treat the Chinese case as merely part of a global pattern of adult concern about the Internet's allure for children and young people (Golub and Lingley, 2008; Tsui, 2005). Notwithstanding the controversy over whether heavy Internet use should be defined as a mental disorder, the notion that Internet addiction can be crippling with regard to work, school, social life and health is a widely shared one in the world. But few other countries have gone quite as far as China in embracing the theory and mounting a public crusade against Internet addiction. According to Potter and Wetherell (1987, 1995), interpretative repertoires are inseparably bound together with larger cultural models of making accounts. A closer look shows that these dual repertoires reflect the ambivalence about the Internet widely found in China, which is co-produced with larger social and political issues under the conditions of 'dual modernity'. The norm about a 'proper' relationship with the Internet expected of young people thus reflects the ideal personhood enjoined of the Chinese subject by the double subjectification regime as discussed in Chapter 1. Such a regime defines the ideal modern Chinese subject as one who is able to exert self-control and is responsible for both one's material well-being and moral order in line with the communist-collectivist values. Moreover, the norm about a proper relationship with the Internet coincides with parental expectations in a context where the notions about success and 'the good life' have changed. As such, any orientation toward the Internet (and other things) that is not conducive to these goals would cause concern among both the Party leaders and the general public, particularly parents, although with differing focuses.

Recognizing that China's ongoing subjectification technologies in the reform era do affect young people's subjectivities and identity construction does not mean to say that resistance (and even rejection) is not possible. According to Rose (1996b: 140), regimes of subjectification should not be taken as deterministic. Resistance to a form of personhood that one is enjoined to adopt is made possible because people live their lives in a constant movement across different human technologies that address them in different ways. Moreover, resistance is necessary not least because the rigidity of the exemplary norms and standards may not prove to be flexible enough to accommodate the complexities of human living, where the practices of everyday life open up numerous spaces of resistance (Bakken, 2000: 345). As I shall show in the following chapter, where I explore urban youth and the Internet café, Chinese urban youth do show their resistance to authorities (parents, the school and society) in ways that they deem meaningful themselves.

6 Between Demonization and Celebration

Chinese Urban Youth and the Net Café

For many Chinese urban youth, the Internet café, *wangba* in Chinese, remains popular despite extensive broadband access at home and despite tightened regulation, constant crackdowns and unfavorable media depictions. What does the *wangba* offer its goers in the Chinese context then? This is the central question to be explored in this chapter. Such an endeavor is much needed, given that whereas the Internet in China has received considerable academic attention, research on the Internet café has been very scarce. This is especially true with regard to *wangba* related perceptions and experiences of young people themselves. The few existing studies have focused on regulatory measures (e.g., Qiu and Zhou, 2005).

HAUNTED BY THE WANGBA

It is hardly possible to talk about youth and the Internet in China without mentioning the Internet café, *wangba*. If the Internet has caused widely shared concern in Chinese society, its emergence in combination with the café as a special social space has made the concern even worse. For many urban parents, the *wangba* appears like a dark shadow over their hearts or a lurking monster ready to spring upon their children. The fear that their children may be lured to the *wangba* haunts them day and night like a demon that can bring misfortune to the family and its cohesion. The image of a desperate and worn-out mother tracking down her teenage child (usually a boy) at the *wangba* and trying to take him back to school is a familiar and worrying one in China. One hears about such stories via the media and one runs into such instances in daily life.

During my fieldtrip in China in 2007, a middle-aged woman contacted me. She wanted to talk with me about her son, a 17-year-old senior high school student, who was playing truant just to 'steep himself in the *wangba*'. We met one afternoon while she was on her way to track him down at one of the hundreds of the *wangbas* in the city. Haggard-looking and with tears in her eyes, she told me that her son was a very good student at primary and junior high school. Seeing the potential for academic achievement in him, about one year before she and her husband decided that the mother would move with the son to the city from a county seat (a much smaller and less developed place) so that her son could go to a better senior high school to prepare for the National College Entrance Exam. They rented an apartment in the city, where the mother served as the caretaker and supervisor of the son. I learned during my fieldtrips that many families resort to such a strategy nowadays as an investment in their children's education. However, to her great disappointment, the boy was getting less and less interested in school and at the time of our conversation, he was constantly absent from school and stayed all day long at the *wangba* instead. It became a daily task for the mother then to hunt him from *wangba* to *wangba*. She said that the boy simply refused to talk at home, shutting the door to his bedroom behind him once he was back at home. He looked depressed and closed off, the mother told me. She wished that I could find out what was wrong with the boy and help 'rectify' him. This episode bears great similarities to the description of a mother and her son offered by the anti-Internet crusader Zhang Chunliang, who claimed that this incident spurred his interest in fighting Internet addiction:

> In September 2004, I saw what was taking place below: A mother in her fifties entered the *wangba* with a despairing look in her face. She walked over to her son, who was totally absorbed in the game on the screen, shouting over and over, 'I'll kill you! I'll kill you!' He was not aware of his mother's arrival at all. Realizing this, the mother threw herself on her knees before her son, saying 'Son, Mama is begging you, let's go to class. Mama has been looking for you for a week. . . . ' Having said this she burst into terrible sobs. The sound of her sobs was heartbreaking.[1]

No one knows how many Chinese youngsters have broken their parents' hearts in this way. Although not all youngsters have gone that far in being obsessed with playing at the Internet café, the worry that their children may become so is real among most Chinese urban parents, as one can easily find out by talking to parents. It is a widely shared perception among parents that the *wangba* represents bad influence on the young, or as one of the parents I chatted with put it emphatically, 'The *wangba* is not by any means a good thing.' The bad effect is not merely in terms of one's worsening school performance due to obsession with playing at the *wangba*. Worse still, bad school performance, according to the large numbers of Chinese parents I have interacted with, always goes hand in hand with a moral deterioration in the young (such as being rebellious to parents and teachers, getting involved with the other sex at an early age, dressing in unconventional ways, smoking, drinking and spending money extravagantly). The notion is: a really good child seldom, if ever, goes to the *wangba*. Moreover, as the *wangba* is known for being crowded, poorly ventilated, smoky and attracting 'unruly' youth, parents also worry about their children's health and safety. In short, the loss that could be induced by a child's *wangba* going is multifaceted. The frustration shared by many parents is thus substantial. An angry father lamented how due to *wangba* going, his 16-year-old daughter, who had been a top student at school, ended up by dropping out. He said with great disgust, 'I wish I could smash all the *wangba*s! Who was so evil as to have invented such a damn thing?' It is typical that parents, awaiting their children back from school, especially from the evening review class, get anxious because they are not sure—and it is not always easy to find out—if the child has actually gone to school or the *wangba*. This is a concern uttered by those Chinese parents I regularly call via Skype or the phone for a chat, as well as those I interacted with in China.

What I have seen and heard is a mere small fraction of the nationwide pool of such stories, with new ones being added to it daily. And yet it is indicative of the far-reaching concerns in Chinese society about the bad influence of the *wangba* on the young, concerns that are well captured in both the local and national news media (Golub and Lingley, 2008; K. Liu, 2004). The *wangba*, thus, has been a central site of governmental regulation (Qiu and Zhou, 2005). This has not only to do with the political dynamics of China's censorship and control of the Internet, as Western media typically propose, but also with the serious concerns of parents, who call for strict regulations and management through local newspapers, television channels and the Internet (Golub and Lingley, 2008; K. Liu, 2004; Qiu, 2003). When a fire broke out in an Internet café in suburban Beijing in June 2002, killing twenty-four young people and injuring thirteen, the government tightened up controls and banned many illegal Internet cafés. Like the more general anxiety about the Internet, underneath the daily and ongoing discussions of the *wangba* in China is the widely shared parental concern about the Internet's distractive effect on the young from school work. Moreover, this goes hand in hand with concern and uncertainty over the direction of the urban youth culture, and of Chinese culture in general, hence the moral order of Chinese society (K. Liu, 2004). Seeking to 'protect' youth, the government has attempted to prevent people under 16 from entering the Internet café (Golub and Lingley, 2008). *Wangba* in some provinces are required to use swipe cards linked to national identity cards, and install surveillance cameras atop computer monitors (Lo, 2009). (However, as is indicated by my informants, this is hardly effective in reality).

This constitutes a striking contrast with the case of some Western societies, where the Internet café has often been lauded as an important alternative initiative for democratizing e-access (Liff and Steward, 2001, 2003; Stewart, 2000; Uotinen, 2003; Wakeford, 1999), and related to this, for the 'new literacy' (Armstrong and Warlick, 2004; Selber, 2004; Street, 2003) and informal learning (Beavis, Nixon and Atkinson, 2005), especially for the 'have-not' social groups. As such, it has been an area for substantial governmental support. In China, on the contrary, the *wangba* has been to a great extent a site that has attracted governmental problematization, interference and restriction, despite its actual role as an effective diffuser of 'e-literacy' in the country's pronounced informatization project (Zhao, 2007).

Despite all the negative associations the *wangba* invokes, its popularity seems to have persisted among urban youth. The Internet café has been showing a robust growth alongside the rapid diffusion of home access (Qiu, 2003). The proportion of café users as part of China's total Internet population soared from 3% in January1999 to 21% in January 2001. Since then, despite tightened governmental regulation, the growth remained relatively stable between 15 and 20% (Qiu and Zhou, 2005). By February 2003, there were approximately 110,000 Internet cafés in the country (not including the thousands of 'underground' ones) (Qiu and Zhou, 2005). In 2007, the number of people surfing at Internet cafés saw a corresponding growth of 60.9% as compared with 2006. CNNIC survey reports in recent years indicate that the proportion remains at above one third, despite nearly universal home broadband access in Chinese cities, for example, with the percentage of netizens accessing the Internet at the *wangba* being respectively 42.4% by the end of 2008 and 35.1% by the end of 2009 (see CNNIC, 2009 and 2010). It seems hard to predict long-term development of the *wangba* given the rapidly changing technological landscape in China. However, at least so far, the *wangba* constitutes an important site for Chinese young people and its attraction cannot be solely attributed to lack of Internet access and availability at home (Qiu and Zhou, 2005). Hanging out in *wangba* has become a popular pastime among the younger generation in many major Chinese cities, although the number of those going online at home has gradually risen over the year (e.g., Lo, 2009; Qiu, 2003; Qiu and Zhou, 2005). According to the report by Niko Partners (2010), at least 65% of gamers who have home Internet access use Internet cafés at least part of the time. The primary reasons are to be social with their friends and to participate in gaming competitions. This attraction of the *wangba* is confirmed by the participants in this study, as indicated in Chapters 4–5. One thing that struck me during my field research in China was that in talking about their Internet experiences, my informants seldom went without mentioning the *wangba*. For many of them, their Internet experience seemed to be closely interwoven with their relationship with the *wangba*.

What lies behind this phenomenon? Of course, this has to do with the relentless efforts of Internet service providers to create a market of mass consumption in line with Chinese urban youth's intense consumerism, with a pursuit of the 'cool' culture, as discussed in Chapter 3. Going online at the Internet café is one form of such consumption. Nevertheless, this explanation is not sufficient. The coexistence of the 'negative portrayal', parental anxiety and strict official control of the *wangba* on the one hand and its persisting popularity among young users on the other is intriguing. There is also the coexistence of improved access at home and young people's continuous preference to go online at the *wangba*. This means that the role the *wangba* plays in urban Chinese young people's everyday lives has become so significant that it constitutes a topic of its own right that deserves close investigation. In other words, the *wangba* is best understood as a special social space of its own right, as has also been argued by researchers studying the Internet café in some Western societies (e.g., see Liff and Steward, 2003; Liff and Lægran, 2003; Lægran, 2002; Lægran and Stewart, 2003). However, there is no such a thing as a universal Internet café. Rather, the Internet café experience can vary from case to case, which has to be understood within the specific socio-cultural context within which the goer's life is embedded.

Drawing on the notion of the Internet café as a 'heterotopian third place', I explore from the *wangba* goers' perspective what Internet café going is all about in the Chinese context. I focus on two questions: What kind of space does the *wangba* offer its users? What perceptions and experiences do the young people have in relation to the *wangba*?

The analysis draws on the three-weeks' observation at a *wangba* (called *wangba* X here) and in-depth interviews with 30 *wangba* goers in the aforementioned city in 2007 (see Methodological Note in the Introduction). The main focus is on the 30 recruited at this site. In addition, I shall occasionally refer to interviews with participants presented in Chapters 4–5 since many of them were also *wangba* goers. Besides, valuable information also came from informal conversations with *wangba* managers and employees.

THE NET CAFÉ AS A 'HETEROTOPIAN THIRD PLACE'

Liff and Steward (2003) have provided a conceptualization of the Internet café as both a heterotopian site from a Foucauldian perspective and a 'third place' from Oldenburg's (1999) point of view. Adapting their line of thinking to my purpose, I conceptualize the Internet café as a 'heterotopian third place', by integrating the two theoretical perspectives of social sites.

According to Foucault (1986), the traditional structuralist notion of historical 'time' is no longer sufficient for a better understanding of our contemporary society. Instead, we need to go beyond the lineal-historical paradigm and view the present as 'above all the epoch of space' (ibid.: 22). He insists that we are at a moment 'when our experience of the world is less that of a long life developing through time than that of a network that connects points and intersects with its own skein' (ibid.: 22). Social space is made up of a multiplicity of different human 'sites', each of which may be seen as a particular 'network of relations'. These sites are 'irreducible' and 'not superimposable', thus turning society into a 'heterogeneous space' (ibid.: 23–24).

Among all these sites, one special type that has drawn Foucault's attention is what he named the 'heterotopia'. Heterotopias 'have the curious property of being in relation with all the other sites, but in such a way as to suspect, neutralize, or invert the set of relations that they happen to designate, mirror, or reflect' (ibid.: 24). They are those spaces which are linked with all the other sites, but meanwhile contradict them. The concept of 'heterotopia', as Foucault has used, thus describes social sites that function in non-hegemonic conditions (Hetherington, 1997). These are spaces of *otherness*, where a certain difference in relation to the hegemonic space of society prevails, and hence where the logic behind the social (re)production of those spaces is necessarily different from the logic of hegemonic space (Casid, 2005). Examples offered by Foucault of established heterotopias range from the cemetery to the cinema. A more recent example can be the social center in the major cities in Europe.

'The heterotopia is capable of juxtaposing in a single real place several spaces, several sites that are in themselves incompatible' (Foucault, 1986). Moreover, 'Heterotopias are most often linked to slices in time', thus breaking with the traditional time (ibid.: 26). Foucault emphasizes that 'the heterotopia begins to function at full capacity when men arrive at a sort of absolute break with their traditional time'. These qualities of the heterotopia create a 'space of illusion that exposes every real space . . . as even more illusory' (ibid.: 27).

Foucault sees the heterotopia not only as a space for the affirmation of difference, but also as a means of escape from authoritarianism and repression. He made this point metaphorically that if we take the boat as the utmost heterotopia, a society without boats is inherently a repressive one: 'In civilizations without boats, dreams dry up, espionage takes the place of adventure, and the police take the place of pirates' (Foucault, 1984: 27).

As Liff and Steward (2003) have suggested, the Foucauldian notion of the heterotopia seems to fit in well with the Internet café as a particular type of social space. The name, a hybrid of the 'Internet' and the 'café', implies a heterotopian juxtaposition of the virtual world of cyberspace and the real social place of the café, two otherwise unrelated or 'incompatible' sites. Drawing on examples in the U.K. and the U.S., Liff and Steward show that the forms that Internet cafés take vary enormously depending upon the 'boundary spanning' between the two elements; but most share the characteristic of being a site where people in a real social setting engage through the world wide web with a 'virtual' arena. This juxtaposition of the real and the virtual, Liff and Steward (2003) propose, forms a profound contrast which appears highly appropriate to the status of a heterotopia, just like the metaphor of the mirror which Foucault applied to illustrate the concept of the heterotopia. The mirror is a heterotopia, a 'placeless place', because it juxtaposes the 'absolutely real' person and space with the 'absolutely unreal' virtual image and its space. This kind of juxtaposition, I would add, takes place not least in the sense that in the 'life on the screen'

(Turkle, 1995) one can have 'another' self, which simultaneously reflects the 'real' person one is and takes upon illusionary characteristics.

Moreover, the sense of illusion also has to do with the fact that the asynchronous features of the Internet enable communication and interaction across the normal constraints of real time, thus breaking with the traditional time. Besides, a break from the 'normal' flow of time also occurs when one is so absorbed in the virtual world, as seems often the case with many young people today, that one forgets about or neglects one's normal routines or tasks in life. The illusionary break with the real time can also take place in role playing games (or other forms of self-refashioning in the 'virtual' world), in which one, or more exactly, the avatar, which is in a sense one's illusionary self, sometimes goes back to the past or goes forward into the future. Such virtual experiences of 'slices of time' from a certain past epoch may be especially true of the Chinese/Asian games based upon ancient legends.

But the sense of illusion made possible by the Internet is not the only quality of the Net café as a heterotopia. The Internet café qualifies as a heterotopia in still another important sense. Cafés, and by extension, Internet cafés, belong to what Oldenburg (1999) calls the 'third places'—'cafés, coffee shops, bookstores, bars, hair salons and other hangouts, that are the 'great good places' at the heart of a community. In contrast to family and work place as the first and second places, such localities serve as central settings of informal public life. Oldenburg (1999: 42) outlined the basic characteristics of third places as follows:

> [They] exist on neutral ground and serve to level their guests to a condition of social equality. Within these places, conversation is the primary activity . . . [third places] are taken for granted and most have a low profile. The character of a third place is determined most of all by its clientele and is marked by a playful mood . . . Though a radically different kind of setting from the home, the third place is remarkably similar to a good home in the psychological comfort and support it extends.

As such, by participating in the activities at a third place, one breaks with the traditional time as one is taking a break from the daily routines. Moreover, with both its distinctive difference from the home and the workplace, *and yet* the spiritually-mentally satisfying effect, the third place both reflects the latter sites and inverts/contradicts them. Thus, a third place serves as a 'harbor', or a boat, to use Foucault's metaphor for the kind of heterotopias we greatly need in today's society, whereby one escapes temporarily from the daily stress of modern life, various forms of authoritarianism and repression. It is in this sense that a 'third place' conceptualized by Oldenburg constitutes a heterotopia in the Foucauldian sense.

In the following, I shall explore empirically how a number of urban Chinese youth perceived and experienced the Internet café as a heterotopian third place. To set the scene further, I shall first offer a brief description of the everyday life at the *wangba*.

EVERYDAY LIFE AT WANGBA X

Since the emergence of the first Internet cafés (or cybercafés or Net cafés) in Europe over a decade ago, this innovative form of e-access has spread all over the world, including many developing countries such as China. However, it is noteworthy that although largely sharing the hybrid name, the Internet café can take on locally specific characteristics. This is true not merely in the sense that the creative combination of Internet and café can take on various forms, differing from café to café (Liff and Steward, 2003; Lægran and Steward, 2003), but also in the sense that neither the Internet nor the Internet café is placeless, but it has to be understood within the specific social-cultural-political context in which it is embedded (Miller and Slater, 2000; Wakeford, 2003).

Wangba X is one of the thousands of Internet cafés that have mushroomed in China's private business sector since the late 1990s. Like many other *wangbas*, it is situated among store fronts such as restaurants, photo shops, bakeries and grocery stores, sharing walls with the adjacent businesses. What distinguishes the *wangba* from the other business fronts are signs with the particular *wangba*'s name in large Chinese characters on the front, featuring 'speed, stimulation and adventure'.

Upon entering *Wangba* X, one was greeted by a *wangguan* (human web manager) at a computer on a small counter by the entrance door. On the wall behind the counter, one could see quite a few licenses and certificates in frames. Next to these were paper posters of official announcements and regulations. On the other walls, there were posters advertising for the most popular (often the latest) games and movies. The computers were arranged in rows in such a way as to best use up the limited space. Users could easily see what the persons seated on either side of them were doing online. There was one ashtray on the desk by each computer (although smoking was not officially allowed there).

Another striking feature (which was also true of all the other *wangbas* in the city) was that there were no such things as writing space, copy machines or printers provided in the physical space of the *wangba,* nor was Microsoft Office software installed on the computers. Only some machines in *Wangba* X had USB slots (whereas in some of the other *wangba*s in the city there were no such functions at all). However, there was a webcam and headphone attached to each and every computer. This was because, some of the managers and *wangguans* explained, no one would have used tools like word processing programs, copy machines or printers and so on even if they had been provided. People were there to entertain themselves by playing games, listening to music, watching movies, chatting and so on, not to work. Each time I entered the *wangba*, I would see most of the users, especially males, who were typically the majority of the visitors, playing games. Some of the patrons were watching movies, listening to music, reading online novels, chatting (sometimes with the Webcam on) and decorating their personal spaces or visiting other people's personal spaces on some social networking sites. I would hear laughing, shouting, swearing among the youngsters while they were interacting with each other both in the physical world and with people (often including their friends present at the *wangba*) in the 'virtual world of games'. The air was often filled with cigarette smoke and other unpleasant smells. All these were similar to what I saw in the other Internet cafés in the city.

The vast majority of users in the *wangba* were young people (mostly males about 15–24 years old). They seldom came alone, but always with at least one friend. Most of those who frequented the *wangba* had membership, which reduces the online fee by 50%. That is, whereas it cost 2 yuan per hour for non-members, it cost only 1 yuan per hour for members. It cost altogether 7 yuan to stay overnight at the *wangba* from 8 p.m. to 7 a.m. It seemed to be the same at the other *wangbas* where I dropped in. On a typical day during busy hours, one could find no vacant seat at the café, but rows of heads with headphones on. I found that it was much easier to get a seat in the morning hours when students were busy at school and migrant workers were at work. I often saw some teenagers or primary school children (mostly boys) rushing into the *wangba* during the lunch break (they came and went running in order to maximize their minutes online). During my fieldwork, I seldom saw a user over 30. I appeared 'exotic' in that context, given my age and purpose of getting online. Thus, despite the fact that they were friendly, I all the time felt that I was somehow out of place.

With regard to food and drinks, *Wangba* X also resembled most of the other ones in the city: typically a *wangba* had a fridge for cold drinks (bottles of water and soft drinks) to sell to the patrons. Some, but not all, sold ready-made snacks such as 'sweet buns', cookies and so on. These things were the same as those one could get from nearby supermarkets. Like most Internet cafés in China, *Wangba* X does not sell coffee, tea or warm snacks. This seems in line with the observation that a typical Internet café in China tends to be predominantly more 'Internet' than café (Hong and Huang, 2005). That is, there is hardly any physical equipment and service typical of a café in the true sense of the word. Whereas this does not necessarily mean that the Internet café fails to serve as a place for sociality (especially given that this may be virtual), as is demonstrated in this study, it does seem to give the Chinese Internet café one of its unique characteristics compared with many of those in the Western countries (see e.g., Lee, 1999; Liff and Steward, 2003; Lægran and Stewart, 2003).

Nevertheless, as Oldenburg (1999) has emphasized, not serving a certain type of beverage or adopting a particular label does not necessarily disqualify a site as a 'third place'. So just as Liff and Steward (2003) have stressed, what is key is whether a place 'works' as a 'third place', rather than what it is called or how it operates. It is fundamentally defined by a distinctive type of sociality and atmosphere, as can be further illustrated by the major themes from the informants' narratives to be presented next.

'WHERE ELSE COULD ONE GO?'

In response to my question 'Why do you like spending your time at the *wangba*?', a predominant response from the informants was the question: 'Where else could one go?' or 'What else could one do?' instead of a concrete explanation. There was only one young man (17-year-old, senior high school student) who came up with a more concrete, albeit still unclear explanation: 'It has become a habit. Whenever we have time, we shall go to the *wangba*. We do not even think.' This taken-for-grantedness aroused my curiosity. I then followed up with a further question: 'But there are so many other places to go and other things to do, such as playing ball games, doing window shopping, going to the park, etc. Why do you think the *wangba* is the only place to go?' Upon this probing, I got a reply from Yu-tao (24-year-old man, vocational school graduate, unemployed), who expressed a shared view:

> Those who often visit the *wangba* include students, some of whom play truant, and people who are older, like me. They do not have any occupation and rely on their family to support them. They do not have much money. They kill (*dafa*) time at the *wangba* as it is both cheap and entertaining.

Like Yu-tao, all the other informants perceived *wangba* going as a cheap consumption affordable for all—at least for the urban families, most of whom have only one child. When I expressed my confusion, 'Cheap? It amounts to much money in the Chinese context if one is online many hours a day and very often', Lin-feng (21-year-old, man, vocational school student), at a group interview with two friends, explained as follows:

> It costs *at most* 20 yuan a day even if one plays 24 hours a day. Even a ticket for the cinema costs 10–20 yuan. Other activities . . . For example, it costs one person 10 yuan to play pooling each hour. Swimming costs much more. Kala OK also costs much more. We are members at the *wangba*. It costs only one yuan per hour (half the normal price) with the membership.

The 'where else to go?' conception was reconfirmed by the *wangba* manager:

> What else can they (children and young people) do? Where else to go? They would have been wandering in the street and causing trouble for their parents without such places to go. People who come here are mostly those who do not like studying or have nothing else to do. Parents would rather give them some money and let them play at the *wangba* than leave them free in the street. Besides, it is hot in summer outside and cold in winter. Inside the Net café, it is comfortable with the air conditioner. And entertaining. It is also the cheapest consumption in the city. So, why not?

As such, the Net café has found its 'niche' among urban children and youth. The widely perceived 'nowhere to go' reminds one of the situation of young people in some other cultural contexts, where they are often excluded from most of (adult) public space (Sibley, 1995; Skelton, 2000; Skelton and Valentine, 1998). Furthermore, as is clear from the previous statements, financial constraints have also contributed to the sense of 'nowhere else to go' in the Chinese case. This points to two interrelated conditions that characterize Chinese city life today. First, there has been a wide-ranging commercialization of leisure activities accompanying the introduction of market economy. As a result, just like in the U.K. context described by Livingstone (2002), but perhaps even more seriously, the commercialization of leisure activities in the reform era has resulted in the shrinking of social space affordable for children and young people, most of whom remain economically dependent upon their parents well into young adulthood due to the pattern of 'emerging adulthood' among the only-child generation (see Chapter 3 and 4). In addition, the reconstruction of the city and urbanization in light of China's pronounced modernization project has been taking up physical spaces previously available for children and young people as free-of-charge and taken-for-granted places to hang out.

Furthermore, marketization of the economy and opening up has fanned up widespread consumerism in Chinese society, especially among young people (see Chapter 1). Such changes impact people's subjectivities in such a way that today what counts as interesting leisure time often carries a monetary tone. That is, it seems that in line with the widespread consumerism, what does not cost money does not count as interesting or entertaining anymore. This is reflected in the 'debate' between five vocational school students and me at a group interview (three of them spoke up while two listened, smiling and nodding their heads):

> *Interviewer:* But there are things that do not cost money or at least cost less.
>
> *A. (22-year-old man):* 'What?' (The others followed him asking 'what?', laughing).

Interviewer: 'Climbing mountains, going to the park, window shopping, just strolling in the street (*guangjie*) with friends, taking a walk and so on. In a word, there are so many alternatives.

B. (22-year-old woman): We do those things occasionally. But one cannot do those things everyday. How boring! One cannot, for example, *guangjie* everyday, can one? (laughing).

A.: Furthermore, one must have money in the pocket strolling in the street (*guangjie*)'.

C. (21-year-old woman): And, it is hot in summer and cold in winter.

Interviewer: When we were at your age, we did all those things. We spent very little money. But we seemed to be no less happy.

A.: (Laughing) But each age has its own form of entertainment/leisure activities. Had we been born some years earlier, we would have felt the same as you did.

Thus, despite their sense of 'nowhere else to go', it would be misleading to assume that the young people meant it as a pure complaint. That *wangba* going is the most affordable consumption in the city is only part of the reasons why it attracts its visitors. As we shall see, in addition to the relatively low barriers to entry, which is often a basic characteristic of 'third places' in Oldenburg's (1999) terms, the participants assigned some other important meanings to this particular site, all of which work together to make the *wangba* a 'third place', which, with its 'heterotopian' qualities that Foucault (1986) has discussed, plays an irreducible role in their lives.

'PEOPLE WHO FREQUENT THE WANGBA BELONG TO THE SAME TYPE.'

One of my questions to the participants and other people I spoke with was: 'What kind of people as you observe often visit the *wangba*?' They would tell me without thinking, 'young people and adolescents'. But most would also point out that most of those who frequent or even 'steep themselves' (*pao*) in the *wangba* are young people who have very little else to do or school students who come during school breaks and those who play truant. In other words, the *wangba* goers see themselves (and were seen by others) as a relatively homogeneous group, different from those who deliberately keep a distance from the *wangba* because of school work or other 'proper' businesses in life. Or as Bi-kong (16-year-old, boy, middle school student) told me, 'People who go to the *wangba* usually belong to the same type.' This is interesting, for despite their shared socio-demographic characteristics (most being the only-child in the family, urbanites, youth, and economically dependent) they formed different age groups ranging from those at school to young adults and they were not all in the same life situation. However, it is the *perception* of belonging to the same type that often helps bring people together to a certain social site. Their sense of belonging is also fostered by the pursuit of a common goal: entertainment and fun. Or as Zheng-liang (15-year-old, boy, middle school student) put it, 'We play what other people usually play (and do). People who go to the *wangba* usually do the same things. They are there to entertain themselves.' That is perhaps why each time I left the *wangba*, the *wangguans* would ask me nicely, 'Won't you play longer, *jie* (elder sister)?' as a greeting and goodbye, even though they knew well that I was not there to 'play'.

Thus, evoking an important feature of 'third places', the *wangba* patrons saw themselves as 'equals'. This sense of being 'equals' seems especially striking when the informants, especially those still at school, contrasted the *wangba* as a social site with the home and school. The *wangba* thus constitutes a heterotopian site for the youth in this study, both reflecting and inverting/contradicting the other sites. Yang-li's (16-year-old, girl, middle school student) words capture this point:

> At school, we are constantly ranked according to our test results. Whereas those who do well become the favourites of the teachers, 'bad' students are despised . . . you often feel that you are good for nothing compared with those good students. At home, you have to listen to your parents all the time. But here at the *wangba*, people are often in a similar situation. Moreover, when we play online, one forgets who one is in the 'real' life. The game offers us a new platform to display (*zhanxian*) ourselves.

As is clear from this example, the 'levelling' function of the *wangba* is not least thanks to the virtual world on the Internet which helps make the *wangba* a heterotopian third place. My informants shared the notion that everyone starts at the same line in the virtual world and is provided with the same opportunity to 'rewrite yourself'. This is in line with the argument by an interviewee mentioned in Chapter 4 that in the virtual world, *everyone* can have another self. Whether such assumptions hold water or not deserves further examination. However, echoing the perceptions about the 'virtual' and 'real' world among the young people presented in Chapter 4, many of the participants in this part of the study experienced a new version of the self enabled by the Net and appreciated the chance to refashion their identities. For example, many (especially men) cited gaining a sense of achievement through online games, which was experienced as much easier than in the 'real' world, as a main reason for being 'addicted' to the Internet. 'It is a good feeling when people admire you and look up to you because you are powerful (*lihai*),' said Lifeng (19-year-old man), who had just completed upper secondary school after striving hard to enter a college during the previous four years. Such experiences seemed to be especially relevant for those who lacked a sense of fulfilment in 'real' life due to 'failure' at school and who, therefore, had disappointed their parents typically with high expectations of their only-children.

It is thanks to the heterotopian characteristics of the Internet that one is able to 'revise', or 're-create' oneself in the virtual world, thus inverting the 'real' world, in which one has been marginalized or excluded. Ironically, however, the 'equals' at the *wangba* may become 'unequal' again in their interaction in the virtual world. Striving for a sense of fulfilment by beating others, they tend to reinforce the notion of competition and stratification in the 'real' world. Nevertheless, my informants emphasized that it is only a game. So, it does not really matter who wins, whereas in the 'real' life it would have mattered greatly. Thus, it seems that the sense of illusion enabled by the Internet in combination with the café is exactly the desired effect of a 'heterotopia', as Liff and Steward (2003) have argued. This kind of 'inversion' and 'contradiction' of the 'real' life by the *wangba* as a heterotopian site may prove especially important for those young people who feel they are marginalized at school or in society in general (e.g., the labor market), which tends to exclude those who have 'failed' academically from important social-economic participations.

ALL FOR THE QIFENG (ATMOSPHERE)

I went to the field expecting that Internet users at the *wangba* did not have access at home. To my surprise, the majority of the young people told me they had home broadband access already. Some come from families who were among the early adopters of the new technology in China in the 1990s. Then, why do they still come to the Internet café, which although affordable for most of the urban only-child families, still amounts to a considerable amount of money for regular users? Some of the parents I spoke with attempted to prevent their children from *wangba* going by offering, although after much hesitation, to have broadband installed at home. However, according to the numerous instances I learned about, such a compromise from the parents' side has not always effective. The child still runs to the *wangba* whenever possible. In addressing the question of why, I learned that one comes to the *wangba* for the atmosphere (*qifen*). Or as Li-tian (24-year-old, employed) said, 'Both my girl friend and I have broadband at home, but we still like to come to the *wangba* (to play). Home lacks the kind of atmosphere (*qifen*).'

At first sight, it was puzzling to me why the *qifen* there should be so attracting. Is it the shouting, the swearing, the laughing and the smoking that permeates the *wangba* that is special? Aren't these exactly some of the elements that have kept many people from patronizing the *wangba*? But my informants were quick to explain that these elements were important for the relaxing and liberating effect of the *wangba*, which was the major attraction. Clearly, they were there for something one hardly found elsewhere: fun, freedom, equality, relaxation and sociability, all of which are interrelated, and which fit the image of the 'third places' that Oldenburg has described. The *wangba* thus served as a temporary escape from the rest of their 'real' life actualities. Such an escape, allowing for a sense of illusion by a break from the traditional time and daily routine, seems especially important for members of the only-child generation, whose lives are typically characterized with great pressure, frustration and boredom.

By the same token, I was told that the closer one was to an exam, the more one was tempted to visit the *wangba*. Particularly, some of the secondary school students said that students go to the *wangba* most frequently when in their last year of school, right before the high school entrance exam (*zhongkao*) and the National College Entrance Exam (*gaokao*). This is in line with what I heard from some of the interviewees presented in Chapter 4. When I asked why one visits the *wangba* more often than usual when one is supposed to be totally bent upon the preparation for exams crucial for one's future, Qifei (18-year-old, man, upper secondary school student) said:

> Oh, the closer one is to the exam time, the more one needs the *wangba*. The reason is simple. We are too stressed. We need to relax in order to cope with the pressure. Your parents and relatives are there expecting a good result from your exams. And you are told time and again how crucial these exams are for your future.

Under such circumstances, one is tempted to escape, albeit just temporarily, into the 'other' world, which in a Foucauldian sense heterotopianly reflects and contradicts/inverts this 'world'. However, given that one can also go online at home or elsewhere, this is not a sufficient explanation. Obviously, this sense of freedom or escape also depends upon the fact that this site serves as a break from the surveillance of authorities such as parents and teachers. It was a shared perception that going online at home was not *guoyin* (thoroughly gratifying) because parents' eyes were always watching them.

As already indicated, besides the feeling of 'freedom and relaxation' hard to come by elsewhere, the kind of sociability one finds at the café is also essential to the '*qifeng*' that has attracted the young people. I seldom saw a young patron come in alone. I was also told at the interviews that they seldom visit the *wangba* alone, but almost always together with at least one friend. Coming from an environment in which people are usually supposed to be quiet when they are online at a public terminal, I was shocked by what I experienced at the *wangba*. Typically the users talked loudly with each other, joking, swearing or giving suggestions as to how to move in the game while they were deeply engaged in virtual life. Obviously, at the *wangba*, it was not a problem for people to talk with each other loudly and freely; on the contrary, this was expected *wangba* behavior. Thus, clearly, both the physical company of friends and the interaction in the virtual world were important for the *wangba* experience. This also seems true of the young people in some other societies (e.g., Beavis, Nixon and Atkinson, 2005).

Moreover, the *wangba* constitutes a 'third place' not only because it is a place for friends to hang out together, but also because it is a place for people to meet and become acquaintances or even friends. In this sense, like the Internet cafés Liff and Steward (2003) studied in the U.K. and the U.S., the *wangba* serves as a social site where both 'strong' and 'weak' ties can develop. This characteristic appears especially important for urban adolescents and young adults given their singleton status, the dismantling of the economic basis for collectivism and comradeship, and the general fragmentation of social life in the era of reform and opening up. This is exactly what Han-tian (23-year-old, unemployed) told me, insisting that today it is not unusual that one does not even know one's neighbor. He told me that he visited the *wangba* not only to interact with people online, but also to meet people in real life.

In short, in pursuit of something that was important but hard to get elsewhere, the young people actively participated in turning the *wangba* into a 'heterotopian third place', which inverts and/contradicts some of the other sites important for them. As such, the young people's trajectory between the *wangba* and the other sites was not always smooth; rather, it was characterized with conflict and negotiation, as I shall show in the following.

SHALL WE 'BIE' TONIGHT?

It is exactly because this particular site contradicts and inverts some of the other sites that there is a tension between what the *wangba* represents and what the latter stand for. Such a tension is aptly captured in Wu-fei's (16-year-old, girl, middle school student) words:

> People's impression of the *wangba* is still very negative. If your mom asks you where you have been and you answer 'I have been to the library', she will be so happy that her face lights up and the two corners of her mouth turn upwards. But if you say you have been to the *wangba*, her face will turn down: 'You are not learning to be good (*xuehao*). How can you visit such places? Is there nowhere else to go?' Scolding, or even beating, will follow.

With this tension, children and young people who are attracted to the *wangb*a have to negotiate with the various types of authorities that are largely skeptical about the *wangba*. Furthermore, parents and teachers may be so strict that the negotiation often involves deceptive strategies. For example, a friend in China told me in 2008 that her daughter, at the age of 16, a high school student then, found her way to the *wangba* every night no matter how the parents and grandparents tried to watch over her. They suspected that she either slipped out of the locked door while all the others were asleep, or, more likely, jumped out of the window from the second floor, where their apartment was located, at the risk of hurting herself. Once, after repeatedly calling her up from bed for breakfast without any response from her, the mother rushed into the girl's bedroom, only to find two pillows under her quilt, meant to give the impression that someone was sleeping there. I also heard time and again from Chinese parents and grandparents that their youngsters stole money in order to play (games) at the *wangba*. A boy's (junior high school student) grandma told me that her grandson once stole 2,000 yuan from her and slipped out at night to play at the *wangba*. Out of fear that she might cause the boy a good beating from his parents and that the anger and frustration upon learning about their child's behavior would have a harmful effect upon their health if she reported to them, she decided to keep it a secret from them.

Similarly, when parental surveillance reached as far as the *wangba*, one often has to lie in order to avoid reprimand or conflict. I heard not only once at the *wangba* some teenagers tell their parents on the phone when the latter called that they were somewhere else doing something 'proper', such as studying, reading, discussing with friends a math problem and so on, rather than at the *wangba*. For instance, I heard a boy telling his mom in a very nice way upon receiving a phone call from her, 'Oh. Mom. Er. . . . I am together with so-and-so. We are trying to solve a math problem at his home. I will go home as soon as we have found a solution'. As soon as he finished the call, he and his friends started laughing.

Such negotiation sometimes takes more creative forms. For example, some of my middle school informants confided to me that the students at their school had to use special expressions that could only be understood among themselves when talking about the *wangba*. They went on to tell me that they never used the term '*wangba*' in front of their teachers, but they used '*bie*' instead. '*Bie*' in Chinese means the turtle, which is colloquially called '*wangba*', two characters that sound very similar to *wangba*, but with a different intonation. Therefore, seeking company to the *wangba*, one would ask the potential companions, 'Shall we '*bie*' (or shall we go to the '*bie*') today?'.

Deceptive resistance was also prevalent among the young adult informants. Although school-age teenagers were the major concern among parents and teachers concerning *wangba* going, the young adults who were 'steeping themselves in the *wangba*' also constituted a major worry for the parents. These older informants, especially the vocational school students, who spent much time at the *wangba*, said that they were doing that against their parents' wishes, so they sometimes had to do that stealthily. For example, Zhaohui (21-year-old, man, vocational student) told me, 'If our parents know we are at the *wangba* so much, they would have pulled us out the moment we sit down and given us a good beating.' Similarly, Lin-tao (22-year-old, unemployed) indicated that his parents had always been against his visiting the *wangba*, but there was very little they could do to stop him, partly because it was hard for them to find out how long and when he was there.

To sum up, even though the tension between this particular site and the other sites made the stay at the *wangba* not always a smooth and nice experience, the young people still valued this site for the qualities it represents which were hard to find elsewhere. It was exactly because of these meanings they assigned to the *wangba* that they saw it as a something indispensable, insisting that even when every home has broadband in China, the *wangba* will still have its niche in society. In response to my question, 'Do you think the *wangba* will disappear when most people have got Internet access at home?', Li-yang (24-year-old, employed) offered another shared opinion among the informants, replying emphatically, 'No, no. It will definitely not happen. It is different to go online at home.'

CONCLUSION

There seems to be a close relationship between many Chinese urban youth and the *wangba* although home access is common and despite the view by both the authorities and lay people that the site is a den of iniquity. This seemingly paradoxical phenomenon has received very little academic attention so far. Thus, we lack an understanding of what the Internet café means for its goers in the Chinese context. In this chapter, I have explored what *wangba* going is all about for urban youth of the only-child generation. The notion of 'heterotopian third place', constructed from Foucault's notion of the 'heterotopia' and Oldenburg's notion of the 'third place' has proven useful for investigating what the *wangba* offers those who partake in its practices—the *wangba* goers—in the form of spaces, in which they can express themselves and feel affiliation with others, forging and re-forging their self and group identities.

The major themes that have emerged indicate that the *wangba* serves as a 'heterotopian third place', or a 'great good place' in Oldenburg's term, in the true sense of the word for its goers. This has to do not merely with the Internet café's inherent heterotopian features in the hybridity of the Net and the café. It also has to do with the young people actively participating in constructing the site into such a space for themselves as they interacted with each other both physically and in the virtual world. That is, neither the computer nor the Internet is interesting in itself; it becomes interesting only when related to the human actors within this particular site (Lægran and Stewart, 2003). As we hear them saying there is nowhere else to go in the city, we are tempted to take their words as a mere complaint about the shrinking of the social and physical space for children and young people today. However, this is only part of the story. The *wangba* represents some unique qualities for them that can hardly be found elsewhere, which qualifies it as a heterotopia with third place characteristics. Such a social space both relates to and inverts and/or contradicts other sites in society, such as the home and the school, which in this case stand in sharp contrast and tension with the *wangba*. Young people come to the *wangba* for something that is lacking in those other sites, from which one needs to escape from time to time. They are there for a sense of being equal, something that is hard to come by in today's China characterized by sharp social stratification, fierce competition, lack of security, consumerism, corruption and unfairness in the distribution of resources. Moreover, the particular atmosphere at the *wangba* turns it into an ideal resort for these urban youth in the midst of pressure, boredom and frustration. By escaping to the *wangba*, even when this has to involve lying and tricks and occur at the cost of one's academic performance, they secure a break from the daily routine to have a sense of illusion in the 'virtual' world and to refashion a self there.

My informants' narratives seem to reflect the shared experience by large numbers of youth who are excluded from the adult culture and who, therefore, need a place to go to get a kind of 'breathing space' (Lægran, 2002: 163). Nevertheless, it is meanwhile locally specific in that their relationship with this particular site is framed by their social-biographical situations shaped by both the personal (such as their age and only-child status) and the general socio-cultural context of today's China. From a Schutzian perspective, it seems that the *wangba*, despite its problematic status in China, has entered the 'zone of primary relevance' in the everyday lifeworlds of many urban youth.

Indulging oneself (as some young people do) in the sense of illusion, made possible by the *wangba*, can be harmful for one's physical and mental health and it may deprive one of the opportunities to participate in other meaningful activities in life. It may also be unwise in a culture where success in life is still largely defined by one's academic achievement assessed mainly through paper-and-pen exams. However, it is important to recognize that young people's celebration of the *wangba* as a heterotopian third place for themselves represents strategies to resist, cope with and subvert authoritarianism and repression in today's China.

In Chapters 4–5 and the present chapter I have explored how urban youth are taking up the new medium (including the new techno-social space of the *wangba*), making it fit into their daily lives while negotiating a 'proper' way to relate to the Internet in line with the moral order of Chinese society, albeit not without resistance. The remaining two empirical chapters are two case studies of how Chinese youth negotiate identities by making use of the discursive space enabled by the Net. Many possibilities exist on the Chinese Internet that could have served this purpose. I choose to look at Chinese youth's online community building and their online political participation. This shows that despite their predominantly entertainment-oriented relationship with the Internet, Chinese youth do employ the Internet for more 'serious' purposes.

7 The Balinghou's Collective Narrative in an Online Forum

Born 1980–1989, they are China's first cohort of only-children due to the one-child policy, the cohort whose journey from the cradle to adulthood has been accompanied by China's initiation and deepening of its reform and opening up, and China's first adopters of the Internet. This special cohort has been referred to as *balinghou*, the post-80ers, who have been at the center of the debate about contemporary Chinese youth. In this chapter, I explore how members of the *balinghou* represent a collective self online.

THE BALINGHOU IN THE SPOTLIGHT

The term '*balinghou*' was first used in media discussions about the young writers born after the year 1980, such as Hanhan and Chunshu, and the cultural phenomenon they represent. It soon became a label for all those born 1980–1989, although in a broader sense, the term also refers to young people born after the year 1980, namely the reform-era generation. Later on, other labels for sub-grouping contemporary Chinese youth (and other age groups) also emerged, such as *qilinghou*, the post-70ers (those born 1970–1979) and *jiulinghou*, the post-90ers (those born 1990–1999). This lends support to Mannheim's (1952/1923) observation that in times of rapid social change, generational consciousness develops more rapidly than during times of more gradual social change. As a result of the spectacular social transformation in the reform era, every decade, even every few years, may represent a dramatically different context for young people in their formative years. This underlies the tendency to label youth by the decade in present-day China.

It seems that none of the other such labels carries such rich and complex connotations as '*balinghou*'. This is not only because of the post-80ers' more direct relationship with a dramatic turning point in Chinese history compared with the other cohorts, but also, and related to this, the controversial nature of their generation. The popularization of the term '*balinghou*' grew as the debate about contemporary Chinese youth occurred in both the media and the wider society. In this debate, the term has become a means of expressing widely shared concern among parents, educators and state leaders about the 'quality' of the younger generation. As Rosen (2009: 360) observes, up until mid-2008, it was common to find youth under attack in the Chinese media, where they were criticized for being selfish, 'reliant and rebellious, cynical and pragmatic, self-centered and equality-obsessed', as well as 'China's first generation of couch potatoes, addicts of online games, patrons of fast food chains, and loyal audiences of Hollywood movies.' They have been referred to as 'the most self-centered generation', 'the me-generation', 'the collapsed generation', 'the generation with no sense of responsibility', 'the strawberry generation' (good-looking but weak-willed and cannot tolerate any pressure or difficulties), 'the generation with no beliefs' and so on. Although these criticisms were directed at present-day Chinese urban youth in general, the *balinghou* seem to have been the main target and have borne the main weight of the attack so far, given that much of such criticism emerged before the post-90ers (born 1990–1999) were born or old enough to participate in youth culture. The Chinese writer, Wang Shuo, publicly called the *balinghou* (writers) 'grandsons' (*sunzi*, a colloquial Chinese term used to belittle people with the connotations of 'being of low social status', 'cowardly' or 'not worth much'), who are brainwashed by Hong Kong-Taiwan culture. And as we shall see in this chapter, many members of the *balinghou* seem to identify themselves with the target of the earlier-mentioned societal problematization, which has served as a motivation for establishing an online community for themselves.

The negative popular image of the post-80ers had been so widely accepted that when members of this generation, reportedly, behaved unselfishly and heroically in the Wenchun earthquake rescue work (and showed their patriotic support for the Beijing Olympic Games) in mid-2008, they gave the Chinese society a 'happy surprise'. People did not expect that even the post-80ers could conduct themselves for the public good and sacrifice themselves for other people's sake. And yet, the change was drastic as well as sudden:

> The same media outlets that had written off such youth now reversed themselves to extol their virtues, while noting, not just in passing, that their altruistic behavior was not surprising because they had learned the virtues of 'great compassion, benevolence, and gallantness' from imbibing traditional Chinese culture, and that, after all, they had 'fully enjoyed the achievements of China's 30 years of reform and opening up.' (Rosen, 2009: 360)

It seems that public debate has started to re-evaluate this generation. However, deep-rooted stereotypes are not to be removed overnight. The debate about the younger generation goes on.

Concern about the collapse of a whole generation may not be new, neither is it found only in China. However, the grounds upon which the concern is based may vary according to historical periods and contexts. The problematization of the post-80s generation in China reflects the widely shared concern about the only-child generation. Ever since the implementation of the one-child policy in 1979, scholarly discussion has worked in tandem with the media and society at large to portray the only-child as the 'spoiled brat', who is egocentric, dependent, weak-willed, lazy, uncooperative, unaffiliative, maladjusted, bad-tempered and disrespectful (e.g., see Liu, 2002a for a review of literature on the Chinese only-child; see also Liu, 2006a). The only-child family has been often described in pathological terms such as 'the only-child syndrome' or the '4–2-1 syndrome'—four doting grandparents, two overindulgent parents, all investing their love, hopes and ambitions in the one child (Wang and Zhai, 1999). The title 'little Emperor' (and by extension, 'little Princess') has often been used to describe the only-child. This has caused a 'moral panic' which sometimes goes beyond China. Some Western observers lament in concord with the Chinese society that the one-child policy has produced 'an emerging generation of spoiled, lazy, selfish, self-centered and overweight children' (e.g., Deane, 1992: 216).

Meanwhile, however, the problematization of this generation has also to do with Chinese social engineers' worry about the 'quality' (*suzhi*) of the Chinese population, which according to them should be improved in line with the double subjectification regime in order to harness both the hopes and risks of modernity, as shown in Chapter 1. To cultivate the desired human qualities, exemplary norms or standards of human quality were carefully tailored out for the population, particularly for the younger generation. Ironically, it is exactly the younger generation that has been often associated with 'the decadence of human quality'. It is obvious that the earlier-mentioned negative generational image of the post-80ers runs directly contrary to these ideals for human beings.

Whenever a group is problematized, its members may feel it necessary to negotiate their collective identity (Mansbridge and Morris, 2001). Or as Hall (1987, 1988) shows, marginal social groups may adopt various creative means to deconstruct and subvert the hegemonic regime of representation by the powerful. In the case of Chinese youth, who typically lack a voice due to their junior status in a society characterized by a patriarchal power hierarchy, how would they represent themselves if given a chance—a chance previously rare, but now possible thanks to the Internet?

I explore the type of collective identity work that is going on in an online forum, www.80end.com, constructed by and for the *balinghou*. Just as one can view personal home pages developed by individuals as a form of self-presentation, or construction of the individual self, whereby the web page becomes a managed site for portraying oneself, one can see online collective web sites as arenas for constructing collective identities (Hine, 2000). I focus on two questions: What kind of collective self are they negotiating in the online forum and how do they represent this particular self?

According to the 'About Us' section on the website, it was established on September 12, 2006. The site's bulletin board system (BBS) or discussion forum (http://bbs.80end.com) has become the main focus of interest and the largest online interaction area for its members. By December 30, 2008, it had 40,358 registered members and hosted 423,733 messages, with a daily average publication of over 1,000 message threads. The number of members is increasing by the hundreds every day.

This chapter draws on qualitative content analysis of message threads collected from www.80end.com, where I have played the role of a non-participant observer for ten months (see the methodological note in Introduction). In doing the fieldwork for this study, I was inspired by Hine (2000), who proposes that the Internet, as both a culture in itself and a cultural artifact, can be seen as field sites for ethnographers and media text for textual analysts. It provides a wide range of possibilities for researchers to conduct web content analysis (Herring, 2009). Web content can focus on 'various types of information "contained" in new media documents, including themes, features, links, and exchanges, all of which can communicate meaning' (Herring, 2009: 11).

My analysis focuses on the message threads (and the follow-up comments by the BBS members) containing opinions and topics that are relevant to the post-80ers' self-understanding and self-representation. The major themes that have emerged will be explored after a brief review of literature on online community and collective identity construction.

THE INTERNET, VIRTUAL COMMUNITY AND COLLECTIVE IDENTITY CONSTRUCTION

It is widely acknowledged that the mass media have been central to collective identity formation by allowing people to 'imagine community' across space and time (Anderson, 1991; Calhoun, 1991). According to Craig Calhoun (1991: 111), they 'offer extraordinary potential for furthering the creation of imagined community, both as objects of identification and as objects of antagonism.'

Compared with traditional media, the Internet has made community building more convenient and on a larger scale and density for people across geographical and social boundaries (Appadurai, 1996; Cullinane, 2003). Cyberspace localities, such as BBSs and other online communication forums, are conducive to community building—basically by facilitating many-to-many communication (for a review of literature on the Internet and community building see Jankowski, 2006). Internet or cyberspace in general, though neither a place nor a fixed space in the common sense, has been often talked about in terms of space, place and locality (e.g., Fletcher, 1998; Hardey, 2002; Rheingold, 2000). For many users, the Internet is, as Hardey (2002) suggests, 'just a different space where they may meet others and make use of a vast number of services and resources' or as a space where they can 'interact with other human beings without actually being in physical proximity' (Fletcher, 1998). Or as Stone (1991: 85) put it, the Internet offers 'incontrovertibly social spaces in which people still meet face-to-face, but under new definitions of both "meet" and "face"'. It is widely observed that such online spaces may constitute what Oldenburg (1999) refers to as 'third places', or 'the great good places', such as cafés and beauty parlors, which serve as central settings of informal public life. A public sphere as conceptualized by Habermas (1989) may emerge within such social sites, whereby participants engage in equal interactions and free communication essential for social democratization, community building and collective identity formation.

Online social spaces may thus fill a need that has been all but abandoned in modern societies, where the closeness and social bonding allegedly have been replaced by fragmentation and emotional disconnect (Oldenburg, 1999; Rheingold, 2000; Jones, 1995). In these settings, individuals and collectivities are engaged in reciprocal discursive construction of identity (Baker, 2001; Giese, 2006). Online communities, like offline ones, allow people to talk in terms of both 'I' and 'we'. However, as Frederic Jameson (1986) argues, aptly capturing the dialectics between the individual and the group, 'the telling of the individual story and individual experience ultimately involves the whole laborious process of telling of the collectivity itself', as the notion of 'community' already presupposes the 'insider' status of its members. Likewise, any form of collective identity that may emerge from the negotiation process unavoidably bears on the individual's sense of self. That is, as Newcomb (1950: 297) argues, 'One's own self and one's own group are interdependently perceived. . . . One may regard the self as a part of the group or the group as a part of the self; in either case they are indispensable, and to the individual both are values'. Furthermore, this process involves negotiation of boundaries between the 'we-ness' and the 'other-ness', making it possible for emerging cultural identities (Giese, 2006). Such processes may bear the potential for agency (Jasper and Polletta, 2001; Snow and McAdam, 2000), as the 'shared sense of we and a common cause can be motivating to act together in the interests of (imagined) collectivities' (Giese, 2006: 15).

The potential of the Internet for community building and collective identity construction seems especially valuable for social groups whose identities have been problematized by the hegemonic culture in society and who normally lack a space for voicing their concerns in the offline world. Previous work has paid attention to the use of the Internet by marginalized social groups such as diasporic transnational communities (e.g., Chan, 2005; Franklin, 2003; Georgiou, 2002; Hiller and Franz, 2004; Mitra, 2003; Ong, 2003; Panagakos, 2003; Parham, 2004) and queer communities (e.g., Nip, 2004; Bryson et al., 2006). These studies show that the Internet offers such groups a forum to negotiate identity by engaging in collective narratives or authorships of the group self.

For many younger Chinese, online communication forums such as the numerous BBSs in Chinese cyberspace may fill 'an existing vacuum in the physical world of Chinese cities by offering urban places of a new quality' (Giese, 2006: 13). The value of the Internet, however, may go beyond this physical sense. Urban youth may especially appreciate the Internet's potential as a public sphere within a socio-cultural context where the joint effect of a host of factors leaves inadequate room for individual self-expression and social-political participation. These include, *inter alia*, their singleton status, high parental expectations, the competitive educational and economic systems, the authoritarian political culture and the lasting influence of a hierarchical tradition that sees young people as juniors in both the family and society. Furthermore, a sense of belonging and community may prove especially important in a society undergoing profound individualization whereby collectivist values are being replaced by individualistic norms, and social cohesion is being threatened by fragmentation.

In line with the 'third place metaphor', research shows that the Internet is widely perceived by Chinese users as a meeting place (Giese, 2006; Guo, 2003; Zhang, 2002; Yang, 2009b), as is also verified by this study. Moreover, as hangouts on the Chinese Internet generally boast a rather informal and playful atmosphere, they also seem to be most conducive to discussing and debating sensitive or controversial topics, which is further supported by the possibility to hide one's true identity (Giese, 2006).

Online localities, however, do not exist in a vacuum. Rather, they serve as localities where real people meet for experiencing community and exchanging information, emotions and opinions closely related to their offline social environment and directly reflecting their experiences as real persons. Besides, online experiences may also influence people's perceptions of reality and behavior in physical places—at least potentially. Given the interrelatedness between their online and offline experiences, the www.80end.com serves as a vantage point for us to gain an understanding of the Chinese post-80ers' perceptions and lived experiences of being who they are and who they aspire to be.

FROM REJECTING TO RECLAIMING THE COLLECTIVE SELF: THE WEBSITE WWW.80END.COM

In Chapter 2, I have shown that reflective of the general Chinese society today, the Chinese Internet is dominated by what Sheldon Lu (2000: 146) has described as an emerging 'consumerist postmodernity', characterized by social fragmentation and a focus on consumerism and life style issues. It is 'more a playground for leisure, socializing and commerce than a hotbed of political activism' (Damm, 2007: 290). However, Chinese netizens do occasionally employ the Internet for various more 'serious' matters as well. Among other things, they employ the Internet for the architecting of online sites for intentional community building, making it possible for social groups to engage in collective identity work (Yang, 2009b). The website www.80end.com serves as an example of the enormous numbers of online communities in China (Yang, 2009b). Although this website shares the 'apolitical', 'consumerist postmodern' nature of the many Chinese websites (Damm, 2007), one cannot miss the collective identity work that is going on there.

According to an introductory thread about the website with the title 'About Us', the website was founded by a group of *balinghou* young people out of their 'strong sense of social responsibility'. As members of this generation, it is said, these young people were perplexed by what they saw as 'the frivolousness of youth among the whole group of post-80ers', which left a bad impression on people of the other age groups, who saw the post-80ers as a generation with no sense of responsibility, neither ideals. Thus, seeing the potential of the Internet for identity work, they wanted to establish a website for the self-improvement or 're-formation' of this generation so as to 'change the bad impression of the post-80ers in other people's eyes'. Therefore, they intentionally chose the domain name '80end, which literally means 'to put an end to the post-80ers (phenomenon)' or 'put to end the post-80ers' immaturity or naivety symptoms'. This means that these members of *balinghou* shared the societal view about the 'decadence' or 'collapsedness' of this generation.

However, as the story of the website goes on to tell us, the aim of the website was completely redefined shortly after its establishment, as some incidents enlightened them to the fact that the post-80ers have actually been misunderstood and misrepresented:

> However, just a few months after the website was established, a newspaper was sent to 80end.com from Shenyang, which told about the success stories of Lixiang, Gaoren, Dai Zhikang and Mao Kankan, four boys from the post-80s generation. The accounts of how they succeeded flew in the face of such notions as 'the *balinghou* is a collapsed generation', 'the *balinghou* is a generation with no sense of responsibility' and 'the *balinghou* are just grandsons (*sunzi*)' in Wangshuo's words. It proved that these are meaningless rumors. It has been proved by facts that the *balinghou* have been always on the right track.

The kind of self-denouncing and self-reprimand has been replaced by a celebration and reclaiming of the *balinghou* collective self. So the story goes on:

> Therefore, the definition of 'www.80end.com' had to be changed, as it was no longer meaningful to end the 'naïvity' of the *balinghou*. After a reevaluation, we redefined 80end.com: '80' refers to the post-80s

generation, 'e' refers the 'e-age' (the Internet era), and 'nd' is the short form for 'year'. Furthermore, it was agreed among the administration team of the website that we will keep constructing the website for 80 years so as to walk together all the time with the post-80s generation.

Accordingly, the ultimate goals of the website were redefined as

> To establish an online garden area (*yuandi*) for the *balinghou* that is characterized by mutual help, mutual enjoyment, and mutual communication; To be concerned about the *balinghou*'s living conditions, record the *balinghou*'s life experiences, and construct an online museum of the *balinghou*.

As such, building an online community and engaging in a collective narrative of their own generation constitute the major purposes of the website.

THE 'NOT-COLLAPSED' 'WE': COUNTERING THE NEGATIVE REPRESENTATION OF THE BALINGHOU

Given the interrelatedness between individual and collective identities, a negative image of a group, no matter whether it is based upon the group's actual conduct or merely out of social prejudice, is unfavorable for both the group as a whole and the individuals involved. It is perhaps no surprise then that the negative collective image of the *balinghou* is perplexing to the BBS participants. It is a widely shared understanding among the members that there has been a misrepresentation of the *balinghou*. Therefore, their collective narrative is characterized by attempts at deconstructing the negative stereotype of the *balinghou* as a group, as demonstrated by the many message threads and comments depicting this generation as the opposite to being 'collapsed', having no ideals or values, not hard-working, being weak-willed, lacking a sense of responsibility, being selfish and so on.

They first of all strive to represent a group image of being capable of great accomplishment, which presupposes such qualities as being ambitious, strong-willed, hardworking and responsible. To highlight such an image, they draw heavily on the post-80ers who have been recognized by society as having made remarkable achievement in a certain field, be it literature, sports, business, science and technology, art or politics. These range from Lixiang, the 26-year-old CEO of the website http://www.pcpop.com, to Dong Ciyang, the young woman who reportedly became president of a company at the age of 21, Li Yuchun, the first prize winner in the 'supergirl' singing contest, Yaoming, the basketball player, Hanhan, the novelist and so on—the *balinghou* young people who have become widely covered by the media. There are threads with titles such as the '*balinghou* entrepreneurs', '*balinghou* millionaires', 'The Ten *balinghou* Wealthy People', 'The *balinghou* Computer Geniuses', 'The *balinghou* People's Representatives', and 'The *balinghou* Internet Elite'. These figures have been given a prominent profile in the online community. A sub-section called the '*balinghou* archive' has been established to record '*balinghou* famous people's names and interviews with them.' These are held as the *balinghou*'s representatives, who have been winning honor for the group and who serve as role models for the others. As such, they serve as social, cultural and symbolic resources for the *balinghou* generation.

To reinforce this 'achieving image', the BBS organizers launched a campaign on October 14, 2008 to recruit more 'excellent *balinghou* workers'. The announcement is entitled 'In Search of *balinghou* Elite: Recommend the Excellent *balinghou* Workers You Know', with the following introduction:

> We used to be called the 'collapsed generation' and the generation with the least sense of responsibility. When our representatives such as Hanhan, Lixiang, Dai Zhigang and Dong Siyang emerged, when we fought a beautiful fight in the Wenchuan Earthquake, society started to reassess us *balinghou*. We are a special group born in a special age. . . . In fact, we are a generation that is not commonplace and who will not be content with being commonplace. Yes. We will prove with our action that we are the most excellent generation.

Furthermore, as the collective profile of a certain social group has very much to do with its social-historical role, defining their own position in relation to China's national development becomes an important part of the BBS members' efforts in constructing their generation as the opposite of being 'collapsed'. In line with their perception of the *balinghou* as a generation of hardworking and achieving youth, they see themselves as moving from the margin of society into the center, assuming the role as the central pillars of China's national development, as is shown by the following messages posted on the occasion of the website's second birthday:

> The *balinghou* generation is destined to be the group that is becoming the mainstream group in Chinese society. No matter how the society develops, they will be the beginning of an epoch. Tracking the footprints of the *balinghou*, the website 80end.com has held up the great *balinghou* flag. Our age is being rewritten by this group. On the occasion of the second birthday of 80end.com, may *balinghou* soon become the central force of society, the pillars of our country and the force that lead the epoch.

> The *balinghou* have turned from a group of children full of wilfulness into the central force in Chinese society that is willing to shoulder the responsibilities of social and national development.

The notion of the achieving 'we' that the members have been trying to establish creates an ethos that inspires its members to strive hard and be proud of their own generation. Therefore, it is both about who they are and who they aspire to be. Threads that promote a highly aspiring and achieving profile of the post-80ers have often elicited comments that cheer the *balinghou* as a group. For example, the following are some typical remarks following such threads:

> Great! Come on. Go, *balinghou*. Carry on! *balinghou*.
>
> We *balinghou* are the most wonderful!
>
> I am proud of *balinghou*.
>
> Come on! Go! We.
>
> Come on. Go! We will work hard together.
>
> Go! *balinghou*. Go, 80end.com!
>
> There are hidden dragons and crouching tigers among us *balinghou*.

Such remarks show a sense of solidarity within the group, which is essential for enhancing the sense of community. The organizers also employ other means to reinforce group solidarity. In addition to efforts aimed at encouraging online interaction (such as collective birthday celebration, encouraging networking among the members and so on), there have also been efforts at linking the members offline and enhancing group consciousness. For example, there was an activity organized by the BBS in relation to its two-year anniversary called 'The Memorial Pamphlet of Youth (*Qingchun Jinian Ce*)', which started on September 24, 2008. It calls on the members to enter for a relay of postcards starting with one of the organizers. The instruction goes like this: Each member shall write on a postcard best wishes/blessings for 80end.com and his or her own BBS ID and pass the card to the next member, who shall get a new card and do the same and pass it on to the next one and so it will go on until it reaches the last member who has entered for the activity.

Enhancing group solidarity, like highlighting an achieving image, no doubt helps the post-80ers to deconstruct the negative collective image imposed upon them. By this, they show that they are a generation that is not only accomplishing individual achievements in various fields, but one that is united.

People who are working hard and accomplishing much, however, can still be self-centered with little sense of social responsibility. Therefore, to further counter the negative image imposed upon their generation, the BBS members make efforts to demonstrate that they are not self-centered individualists concerned only about themselves, as they are often accused of. For this purpose, no other evidence could be more telling than the (reportedly) commendable conduct of the post-80ers during the Wenchun earthquake (May 12, 2008). Therefore, the BBS members reverberate with the media to make a strong case of this. This part of their collective narrative builds upon the observation that members of the post-80s generation constituted the main force in the rescue and aid work during the Wenchuan earthquake. This may have to do with the fact that most of the current military members were born in the 1980s, as they form the major recruitment pool for the army. But it was reported that there were also many non-military post-80ers volunteering. According the media, the post-80ers not only took an active role in donating money, materials, equipment and blood but they were also active in the rescue work on the site. There emerged a number of *balinghou* heroes from the earthquake, whose stories have been much covered by the media. The conclusion is that the post-80ers are totally different from the stereotype of being selfish, without a sense of social responsibility, or easily crushed by pressures and difficulties. Thus, the post-80ers' prominent profile during the earthquake not only provides fresh material for the Chinese social engineers to establish more role models for the younger generation, but it also gives the *balinghou* an opportunity to voice their collective claim of being the 'not-collapsed' generation, as one of the BBS members stated, 'We have always been great, but it is only that they did not recognize us'.

Such humanitarian efforts aimed at relieving others of pain no doubt indicate that the post-80ers can sacrifice their own interests for other people when necessary. And, they wish to demonstrate that this not only applies to people unknown to them but also to their own parents. There are some threads talking about the *balinghou*'s willingness to fulfill the filial duty expected of them despite difficulties. The fact that there has not yet been much said about how they are actually fulfilling the filial duty may be due to the 'emerging adulthood' phenomenon among this generation (Nelson and Chen, 2007). However, they do make use of such evidence as exists to show that they are dutiful children to their parents as well as responsible citizens. For instance, there is a thread that highlights a survey done in the West which shows that the post-80ers who are abroad attach great importance to filial duty. The title of the thread goes: 'The post-80s overseas Chinese have grown up and they see filial duty as their first priority'. I also came across remarks in the BBS which sound like declarations to show their determination to fulfil the filial duty, such as: 'Please, the whole society, rest at ease. We *balinghou* will shoulder the responsibility for our parents despite the hardships' and 'We still cherish the traditional values of the filial duty'. The members also actively participated in a debate about a *balinghou* young woman who reportedly served as a prostitute online in order to earn money to save her diseased father's life. Although it was debated as to whether her action was wise, her filial heart was much admired and caused sympathy among the members of the online community, who donated money and gave her advice to help her out. This incident has been entered into the 'milestones of the 80end.com/bbs', a list of all the events and incidents related to their online community that the members deem of great meaning for their collective self or as a strong indication of who they are:

> At the beginning of June 2007, we focused our attention on the 24-year-old girl who 'sold her body' online in order to save her father with heavy

> disease. We established denoting groups for the heroine Huang Meihong and made websites dedicated to this matter. Our electronic journal interviewed her. Such acts are much approved by the netizens nationwide.

According to the BBS participants, another defining 'quality' of the post-80s generation is their love for their motherland. Research shows that nationalism is a form of collectivism widely internalized among Chinese youth today (Zheng, 1999; Zhao, 2004). Patriotism is an indicator that is often used in the assessment of a person's or group's a moral integrity in today's China (see Chapter 8). To demonstrate that the post-80ers are patriotic, the BBS members cite the case of overseas Chinese young people, many of them are *balinghou*, who behaved fearlessly in protecting the 'sacred fire' (*shenghuo*) of the 2008 Olympics when the torch relay was disturbed in some cities in the West. The rescue action after the earthquake has also been referred to as a demonstration of their love for the country, a point made in threads such as 'The *balinghou*'s Chinese Heart' and 'The *balinghou* generation: the beautiful successor of Chinese Culture'.

THE UNIQUE 'WE': CELEBRATING THE MULTI-FACETED SELF

Identity construction is much about claims about 'uniqueness' of the self (Hall, 1997). However, people may employ different strategies trying to achieve this goal. In this case, the post-80ers seem to take pains to demonstrate that their 'uniqueness' is defined by a multi-faceted self, rather than the simplified image assigned to them by society. This lends support to the argument that people tend to perceive their own self as multi-faceted and as more multi-faceted than others (Sande et al., 1988). This may be especially true of cases such as this one, given the *balinghou*'s collective goal of subverting the kind of stereotype imposed upon them. By projecting themselves as unique based upon multi-facetedness, they not only draw the boundary between the 'we' and the 'others', but also defy simplified notions of this generation, particularly those that categorize their 'uniqueness' in terms of shortcomings.

In the ongoing discussion among the BBS members, there is a shared understanding that the *balinghou* are the products of a special epoch, in which three major factors shape the context in which the *balinghou* were born and grew up. These are the only-child policy, the market economy and the Internet, all of which work in tandem with each other to impact upon this generation. As a result, this generation has come to bear characteristics peculiar to themselves, shared neither with predecessors nor with successors. They tend to use terms such as 'the most', 'unprecedented', 'never before', 'the last' and 'the first' in describing themselves. Furthermore, some of the members have constructed lists of *balinghou*'s characteristics to share with the other members, as is, for example, reflected in threads such as 'the 30 characteristics of the *balinghou*', 'life principles that only the *balinghou* understand', 'the 26 living principles for the *balinghou*', 'the *balinghou*'s 40 life philosophical principles', and 'the differences between the post-70ers, the post-80ers and the post-90ers'.

What is striking is that there is a persistent tendency in these lists and many other threads to describe the post-80ers as representing sets of seemingly contradictory qualities, which nevertheless are coherently combined in the post-80ers' multi-faceted personality. The message conveyed is that these seemingly contradictory qualities are mutually complementary and they are all necessary for a more comprehensive understanding of this generation. The following are some of the statements from the message threads in the BBS:

> The post-80s generation is both romantic and realistic at the same time.
>
> Feeling a bit scared whenever we feel happy, shedding tears (because we are moved) whenever we are laughing. This is the *balinghou*.
>
> The *balinghou* are often questioned and criticized, but still optimistically persisting to be themselves and forging forward.
>
> (The *balinghou*) is a group that is full of the Zeitgeist spirit of the new age but meanwhile shouldering the task of carrying on traditions.

Being at the same time the embodiment of the old and the new, as the last quote indicates, is seen by the BBS members as the greatest quality that adds to their uniqueness. It is argued that as the *balinghou* is the first generation that has been born and growing up in the dramatic social transformation of Chinese society, they naturally become the product of the clashes between the old and the new, thus bearing the elements of both. Whereas one may argue that every generation is the bearer of both the old and the new, they make the point that being the *first*, and most likely the *last*, generation in Chinese society that has been simultaneously experiencing all the aforementioned detraditionalizing forces, the *balinghou* seem to experience the clashes more keenly than other generations. Therefore, they see themselves as 'the link between two epochs', in one of the participant's words. This quality defined by their historical position, is further captured in the thread entitled 'the differences between the post-70ers, the post-80ers and the post-90ers'. Among the many differences presented in the list, there are some that go along the same story line of being the bridge between the old and the new:

> The post-70ers: they want to marry women who are virgins.
>
> The post-80ers: That does not matter. As long as we love each other, it is OK.
>
> The post-90ers: Does one need love to marry? Does one need to marry at all?

> The post-70ers: they take off their shoes before entering home.
>
> The post-80ers: We enter home without taking off the shoes.
>
> The post-90ers: They even go to bed without taking off the shoes.
>
> The post-70ers: whenever they see their leaders standing, they will offer the seat at once.
>
> The post-80ers: We advocate equality between leaders and the followers.
>
> The post-90ers: Between heaven and earth, I am the sole authority (*weiwo duzun*).

However, it may not always feel 'cool' to be the 'bridge between the old and new', as the earlier comparisons seem to convey. It is exactly because they are born into a transitional period and serve as the bridge between the old and new that the *balinghou* may at times feel lost and frustrated. Thus, in agreement with the young adult only-children I studied earlier (Liu, 2006a), their collective narrative is characterized by another binary combination: choice and risk. As one of the BBS participants commented, 'Never before was there another group that is simultaneously embracing great wealth and a wide range of opportunities and undergoing the frustration and torturing of feeling lost.' In line with this, some even see their generation as the 'sacrificed generation' in the sense that they have served as unwitting social experiments in China's pronounced modernization project: they are the first cohort of only-children, the products of the only-child policy, the first group to shoulder the filial duty single-handed without the help of any siblings, neither much social security, the first generation growing up with market economy and opening up—in short, the first generation that has been exposed to many new (but not necessarily good) things.

Nevertheless, emerging from their reflections upon both choice and risk is a self that is under great burdens, and yet perseverant; and confronted with hardships, and yet optimistic. Thus, they find another set of simultaneously complementary and contradictory qualities that define their collective character. This is captured by the following remark posted on the occasion of the BBS's two-year birthday by a *balinghou* high professional:

> *Balinghou* is a special term for a thought-provoking social group: The term symbolizes loss. She (the author uses this pronoun to show her love for her own generation) carries expectations, she is full of longing, she wanders about, she seeks far and near, up and down, she endures all tortures, but she persists in her ideals and she forges forward bravely. Bless you, the forever *balinghou*, and the forever 80end.com!

The earlier-mentioned wide range of combinations of seemingly contradictory qualities/features that the post-80ers have assembled to define themselves is extended in a widely circulated posting entitled 'The 26 Living Principles of the *balinghou*', in which each of the 26 letters in the English alphabet represents a word that assigns a defining quality of the *balinghou* generation. In the range of 'qualities', 'values', 'notions' and 'characteristics' assigned to the *balinghou*, this group is described as, and expected to be, simultaneously dreamers and realists, hard workers and fun seekers, the consumerist, who is concerned about environment, the modern citizen who is civilized but may defy discipline, the innovator and fashion chaser who cares about traditional culture, the individualist with a concern about public good and others' well being, showing a sense of social responsibility, the optimist who is under great pressure, challenges and risks and so on. It is hard to see if these are principles by which the *balinghou* are enjoined to live, as the title of the message thread indicates, or claims of qualities that they have already acquired. In other words, it seems again that these are as much about who they are as who they aspire to be.

In short, according to their self descriptions, the post-80ers incorporate both hedonism and 'Puritanism', both individualism and collectivism, both modernism and traditionalism, both romanticism and realism. The series of dual qualities/characteristics that the post-80s members claim they embody together constitute the multi-facetedness which is essential to their claim of uniqueness and essential to their rejection of any simplified portrayal of their generation. For this reason, the BBS members highly celebrate this multi-faceted self. For example, in a thread entitled 'The *balinghou*'s Living and Spiritual Conditions', the author writes, 'Some *balinghou* say they have a double-sided personality. But (I will say that) we *balinghou* are multi-sided. . . . We have inexhaustible power.'

In line with this, the members in the online community urge each other to live out and up to their true self, as is reflected in comments such as:

> In exhortation with all the members of our 80end.com, I suggest that we *balinghou* live up to our true self.
>
> *Balinghou*, persist in being ourselves.
>
> We will do what we want to do.

Furthermore, they seem not content with merely keeping their 'unique' self to themselves. Instead, they wish to pass on or spread the so-called '*balinghou* spirit' in society, as a *balinghou* writer writes:

> The sky of the *balinghou* is full of boundless colors. We are a group that likes to shout. We are an unusual group that is holding up an unrivaled sail in the turbulent waves of this particular age. May the 80end.com become better and better, whereby we drive the *balinghou* production forces into the new epoch and let the songs of the *balinghou* spread everywhere.

Based upon this self perception, they call for society's understanding and tolerance of this generation so as to provide them with more opportunities to develop their potential. It is claimed that those who do not know how to deal with the *balinghou* due to a lack of understanding of this generation from the latter's perspective will lose their future, as is reflected in a thread with the title 'Growing up together with the *balinghou*: When you have learned how to get along with the *balinghou*, you have grasped the skill to cope with the future'.

THE NOSTALGIC 'WE': COLLECTIVE REMEMBERING OF THE GOLDEN CHILDHOOD

One can hardly get about in the BBS without noticing that the members have been carrying on a large-scale collective recalling of their childhood with the help of the medium. A separate section called 'Nostalgia' (*huaijiu*, which means 'missing the past' in Chinese) has been devoted to such activities in the BBS. There has been accumulated a wide range of collections of the so-called 'classic memories of the *balinghou*'s childhood', such as collections of *balinghou* childhood games, toys, nursery songs, doggerel verses, school life episodes (such as lining up after school), often used sentence patterns in Chinese compositions, posters, cartoons, books, textbooks, songs, heroes, model figures, movie stars, singers, films, snacks, clothes, hairstyles, advertisements, jokes, ideals, stories, fairy tales, TV programs and even the tools parents used to beat them. Some of them took pains to collect these things and present them online to share with the others, who may comment upon them and sometimes add more to the collections. These collections typically illicit comments such as:

> Oh. How touching!
>
> I am touched to tears.
>
> I really miss those things!
>
> Oh, how I miss those days!
>
> How I miss my childhood!
>
> Many of the good-to-eat and good-to-play-with things have disappeared.
>
> It is hard to find these things today.
>
> I feel these were so far in the past.
>
> It appears to me that these things were in the 1960s. So long ago.

Nostalgia is often associated with aging and/or diasporic experiences (Brown and Humphreys, 2002), but less with young people. It may therefore seem surprising that the BBS members have displayed a strong tendency for nostalgic feelings. Nevertheless, this may not be a surprise given the role nostalgia plays in individuals' and groups' sense of self (Sedikides and Wildschut, 2008; Sedikides, Wildschut, and Baden, 2004). According to Sedikides and colleagues (ibid.), nostalgia has four key psychological functions: it serves as a repository of positive affect, maintains and increases self-positivity, fosters affiliation or stronger social bonds and carries existential meaning, serving as a reservoir of memories and experiences that is helpful for coping with existential threat. These views are supported by some other studies. For example, in a study by Brown and Humphreys (2002) of the collective identity narrative of a Turkish faculty of vocational education, special attention is paid to the importance of nostalgia in the acts of collective self-authorship. Nostalgia, the authors argue, can be a means of maintaining a collective sense of socio-historic continuity, a source of resistance to hegemonic influence and a defense against anxiety. Likewise, Yang's (2003b) study of the nostalgia widely observed among China's *zhiqing* (urban youth sent to the countryside for 're-education' during the Cultural Revolution, 1966–1976) generation indicates that nostalgia helps the *zhiqing*, who find themselves again in the vortex of social change and in renewed struggles against disruptions of identity in the reform era, to maintain and construct identities by connecting the present to the past, articulating past experiences and their meanings and directing moral critiques at the present.

The *balinghou* members' 'collective remembering' serves several important functions. It constitutes the processes by which the post-80ers as a group assemble shared storylines that subjectively constitute their collective identity. It has been documented that constructing a shared past is an integral part of the general process of negotiating collective identities (Bellah, et al., 2008; Anderson, 1991, Wertsch, 2002). Moreover, nostalgia gives access to a shared childhood that is peculiar to the *balinghou* 'us'. It thus enhances the cohesiveness and affiliation among the members and gives further legitimacy for the group's uniqueness claim. Besides, acts of collective nostalgia such as those carried out by the BBS members can inform the stories by which individuals and groups understand their present circumstances, preserve self-esteem and react to perceived threats. It thus serves as an emotional support in an era of rapid social transformation, where pressure and uncertainty prevail. This seems much needed given the *balinghou* members' voiced feelings of loss and frustration.

This last point seems especially relevant to the BBS members, who in their discussion about why the *balinghou*, albeit still young, have started to be nostalgic, associated this tendency with the fact that they have been under too much pressure. In line with the discussion carried out in Chapter 3, the perception that the pressure on the post-80ers is too great constitutes another striking theme in the BBS discussions. During my visits to the BBS, I often came across utterances with strong emphasis, such as: 'The pressure on us *balinghou* from the family and society is too great!', 'In fact, we *balinghou* are really very exhausted, unlike the post-90ers, who do not need to worry about anything' and 'the post-70ers and post-90ers are much happier than we are'. The pressure, it is maintained, is the combined effect of the lack of security and stability, fierce competition, lack of employment opportunities, the emerging burden of filial duty expected of the only-child, the task of child bearing that is getting more and more current for the *balinghou*, who are not quite ready for such tasks yet, and the changing notions of what constitutes 'the good life'. The BBS members often come up with their personal stories to illustrate the pressure faced by the *balinghou* as a group. For example, in telling about their experiences of the employment situation and professional development for young people in China, some lament that 'today college graduates lose their jobs as soon as they graduate.' A member speaks on behalf of the *balinghou* group as follows:

> For most of the *balinghou*, it is common to start from the scratches. Unlike the post-70ers, who benefited from the reform and the post-90ers who can rely upon their parents, who are often better off than our parents. . . . We have to rely on ourselves to buy an apartment and support our parents.

Given the financial pressure to be incurred in maintaining a family according to the middle-class life style, which has become a norm among China's urban youth, it is debated whether the *balinghou* can have a stable marriage. According to one of the members, the post-80ers' marriage will be greatly affected by the fact that many of them cannot afford to buy an apartment, as the title of the posting indicates, 'At 26 I have got divorced: the *balinghou*'s marriage is doomed to die due to the housing issue'. She attributed her divorce shortly after her marriage to the fact that they did not have an apartment of their own, which led to much instability, hence anxiety for the two of them, who quarreled very much. Some agreed with her and maintained that lacking the financial ability to buy an apartment and other necessities for a marriage and family life according to the new standards is the main cause of the phenomenon that the majority of the *balinghou* put off establishing a family. In a similar way, some lamented, 'The *balinghou* cannot afford to fall in love'. Thus, whether the post-80ers should have a career first or get married first is an ongoing debate within the BBS.

As we can see, just as the BBS serves as a medium for their collective remembering, it also serves as a 'third place' where the young people share their experiences of burdens, hardships and challenges. Such collective commiserations reinforce shared experiences, whereby group members may feel that they are substantially different from outsiders, which is conducive to a shared sense of identity.

Moreover, such collective commiserations hold the key to an understanding of their widely shared nostalgia for a 'not-so-long-ago' past. Thus, against this background of pressure, nostalgia allows them to escape to a time when life was quite different from the present: happy and carefree, as the following quotes further demonstrate:

> This reminds me that even I used to have a happy childhood.

> In fact, we were very happy compared with children today. At least we did not need to go to the after-school English class, did not need to learn so many things. We were really very happy then.

> Although our childhood was relatively poor, although the ice-cream bar cost only one *jiao* (ten Chinese cents), neither *heluxue* nor *menglong* (the two ice cream brands considered most exclusive in China), we still felt happy.
>
> Yes, I remember eating liquor chocolate and soda-water candy. These have disappeared now. Oh, I miss them!
>
> Suddenly I feel so moved (*gankai*). I do not feel I am old yet. Nevertheless, just feel I cannot get back to the days when I had nothing to worry about.

Interestingly, their construction of the past seems to be quite discrepant with the impression one has of the only-child's childhood. It is a well-known phenomenon that due to interrelated factors such as only-child status, great parental expectations, the exam-oriented education, and the fierce competition in Chinese society today, the post-80ers used to be under great pressure to achieve from the very beginning of their lives, as indicated in Chapter 3. This may well be the more realistic picture. Nevertheless, just as Maurice Halbwachs (1950: 69), who first articulated the notion of collective memory, explains, 'a remembrance is in very large measure a reconstruction of the past achieved with data borrowed from the present, a reconstruction prepared, furthermore, by reconstructions of earlier periods wherein past images had already been altered'. That is, with each retelling, an account 'from the past' takes on new form and meaning, which is closely related to one's current life situation. As such, the collective remembering among the *balinghou* not only has to do with their wish to construct the BBS as an online museum of and for themselves, but it also represents a strategy in coping with existential problems. In this sense, like the case of the *zhiqing* described by Yang (2003b), nostalgia serves as a collective cultural resistance in the *balinghou*'s collective narrative in a socio-cultural context characterized with extreme rationalization, materialism, fierce competition, sharp social stratification and fragmentation.

CONCLUSION

The context in which Chinese young people's identity construction is being played out is dramatically different from those of the previous generations due to the profound social transformation in the reform era. The problematization of the post-80s cohort, whose path to adulthood has been accompanied by the dramatic and profound social changes, represents widely shared concerns about meaning and identity under changing conditions of Chinese life. The quest for meaning and identity equally applies to members of this generation themselves, a process that is not only related to society's negative stereotyping of their group identity, but also to the larger socio-cultural context of today's China in which their lives are embedded. The Internet, with its potential to provide virtual localities with 'third place' qualities which usually are hard to come by in China, has proved to be especially important for young people to have a sense of community and to engage in a collective narrative. The exploration in this chapter serves as an attempt to understand how individuals and groups utilize online resources creatively in negotiating their meaning making in an age when the organic unity of the world is no longer taken for granted.

Emerging from the many message threads I have analyzed is a particular collective self that goes far beyond a mere deconstruction or subversion of the stereotype of the post-80s generation. By actively (re)constructing a particular collective narrative in the BBS largely by means of posting message threads, they have written community and a collective self into being (Boyd, 2008). In portraying their collective self as multi-faceted, they not only distinguish themselves from the 'others', which is important for community building, but also defy and reject any forms of simplified understanding of this generation. Their claim of uniqueness and multi-facetedness, as well as group cohesiveness, has been reinforced by the collective remembering of a common past. This process characterized by a strong nostalgia for a 'not-so-long-ago' past represents their reflections about a rapid-changing society which confronts individuals with unprecedented challenges, pressure and frustration. It further reflects their perceptions and experiences of the pressure incurred and to be incurred by a range of interrelated factors, such as only-child status, fierce competition, changed norms about what constitutes 'the good life', lack of security for both the young and the elderly, and the daunting filial duty expected of them as only-children. Nostalgia is not the only means for voicing frustration and pressure. The virtual locality of the website in general, especially the BBS, can serve as a forum for commiserations of shared pains and frustration.

In constructing the post-80s generation as a particular group, the young people's online self-representation is often characterized by a combination of descriptions of who they perceive they are with normative and aspirational enjoinments of who they should become. Thus, it is important to understand their online discussion not only as about their 'being' but also about their 'becoming', that is, it is not merely about who they are but also about who they aspire to become. As such, the post-80s' collective narrative in the online forum lends support to Margret Hall's (1990: 138) argument:

> Identity calls imagination into play—we are what others perceive us to be, and also what we perceive ourselves to be. When we strengthen the idealistic component of identity, we transcend others' perceptions and seize our own destinies. We become more than what others see us to be

> by articulating our ideals in identity. Ideals are key means of transcending our restricted roles and limited goals.

Both who they perceive they are and who they aspire to be seem much in line with the ideal 'human qualities' enjoined by the double regime of subjectification, with a dual emphasis on both an entrepreneurial spirit and moral strength with a strong altruist-collectivist orientation. A closer look, however, shows that their talk about such values is framed largely in a humanitarian understanding rather than the socialist ideology typical of the official discourse. Furthermore, their narrative is meanwhile characterized with attempts to present a self that incorporates 'dual qualities' that go beyond those enjoined of the double subjectification regime. The apparent paradoxes characteristic of Chinese urban youth (discussed in Chapter 3) are thus mobilized as complementary parts of a coherent, albeit multifaceted whole in their own ideal form of self. Such a dual approach to identity construction not only represents youth's responses to the cultural and political conventions embodied in China's 'dual modernity' but also illustrates the view that identity construction is an emergent and ongoing narrative project which is meanwhile characterized by a struggle for coherence (Budgeon, 2003; Goodson and Sikes, 2001; Josselson, 1995, 1996). Like Chinese youth's negotiation of a 'proper' wired self (Chapter 5), the collective narrative analyzed in this chapter is indicative of identity struggles of individuals and groups caught between exemplary norms of human quality and their lived experiences of modernization with Chinese characteristics. Such search for a coherent self within a socio-cultural condition full of apparent paradoxes and contradictions is further reflected in Chinese youth's negotiation of a political self—a topic I will turn to in the following chapter.

8 From Political Indifference to Vehement Nationalism

Chinese Young People Negotiating the Political Self in the Internet Age

Flaring nationalism in Chinese cyberspace, especially in relation to the 2008 Olympics, constitutes a striking contrast to the normally apolitical nature of Chinese netizens' use of the Net. Why do a political indifference and a nationalist passion coexist in Chinese cyberspace? What can this tell us about young people's negotiation of a political self in the socio-cultural context of reform-era China? This chapter explores these questions from the netizens' own perspectives by locating the informants' narratives within a re-conceptualization of political participation, taking into account young people's nationalistic sentiment and behavior in China. Like the preceding chapter, this chapter illustrates how Chinese youth employ the Internet for negotiating 'modern' Chinese identities.

A number of incidents in relation to the 2008 Olympic Games, such as the anti-government protests in Tibet, the disturbances during the international leg of the Olympic torch relay, CNN commentator Jack Cafferty's remarks against the Chinese and the alleged 'anti-Chinese' tendency of CNN and some other Western media sparked a sharp surge of nationalism among Chinese both at home and abroad. This outburst of nationalism spanned the borders between cyberspace and the 'real' world and between Chinese at home and those abroad. Inside China, where political passions are usually kept in check by government control, much of the outrage was vented out in cyberspace, to which at least 210 million Chinese (mostly educated, young people under 35) had access then (CNNIC, 2008). Just to give a few examples, Chinese netizens set up a series of nationalistic websites such as www.ANTI-CNN.com (with a thematic video song entitled '*China's Return to Prosperity Cannot Be Prevented*') and www.dalai-liar.com. A lyric entitled 'Do not be too CNN' composed by netizens and sung by a young man in a rap style accompanied by a video was circulating in Chinese cyberspace. The phrase 'Do not be too CNN' became one of the most popular online expressions in China in 2008. By mid April 2008, a Google search in Chinese for boycott of 'Carrefour', a French chain store, which it was said, was giving financial assistance to the Dalai Lama, yielded over 2.4 million hits, most of them created within the previous week. Some 2.3 million MSN users had attached 'I love China' icons to their online profiles and many MSN users put a 'red heart China' in their screen names as an expression of their solidarity against the 'Tibetan separatists' and the West. A patriotic Internet petition calling on Chinese people from all over the world to oppose 'splittism' and protect the Olympic flame had been supported by more than 8.16 million people on sina.com by July 1. There were many other websites with similar petitions.

Flaring nationalism in Chinese cyberspace is not new (for Chinese cyber nationalism, see Wu, 2007). There had been outbursts of nationalist sentiment both online and offline over issues such as the Japanese occupation of the Diaoyu Islands and the NATO bombing of the Chinese embassy in Belgrade in 1999. China also has a breed of so-called 'red' hackers who pride themselves for sabotaging and defacing websites in countries that are perceived as anti-China, such as Japan, Indonesia and the U.S. (Wu, 2007). However, what is interesting about the surging cyber nationalism is that it constituted a striking contrast with the normally 'apolitical' nature of Chinese netizens' use of the Internet, which is predominantly focused on entertainment, lifestyles and private issues, as shown in Chapter 2.

This chapter explores such a contrast by focusing on two questions: How do young people perceive and experience the Internet as a political tool within the specific context of today's China? How do they understand political participation and nationalism?

It draws on semi-structured interviews with the 38 university students (20 men and 18 women) (see the methodological note in the Introduction). In addition to the interviews, valuable information came from browsing of Chinese websites from mid-March to the end of May 2008. I paid special attention to postings that had to do with the incidents related to the 2008 Olympics.

RE-CONCEPTUALIZING POLITICAL PARTICIPATION

Verba and Nie (1972: 2–3) define political participation as 'acts that aim at influencing the government, either by affecting the choice of government personnel or by affecting the choices made by government personnel.' Many have sought a broader conceptualization of the term (e.g., Barnes and Kaase, 1979; Conge, 1988; Norris, 2002). Seeking to work out a definition that is broad enough to encompass a range of behavior in a variety of cultural settings, Conge (1988: 247) defines political participation as 'individual or collective action at the national or local level that supports or opposes state structures, authorities, and/or decisions regarding allocation of public goods.' Conge further stresses that the action can be verbal or written, violent or non-violent, and it can be of any intensity. This comes close to Norris' (2002: 4) definition of political participation as 'any dimensions of activity that either designed directly to influence government agencies and the policy process, or indirectly to impact civil society, or which attempt to alter systematic patterns of social behaviour.'

Attempts such as those just mentioned at broadening the concept of political participation are especially necessary today, when new understandings and new forms of political engagement have been documented among young people in many societies (Hoskins, 2005; O'Toole, 2003; White, Bruce and Ritchie, 2000). This is especially true in societies where new media seem to have transformed the context in which young people grow up and their identity construction is being played out. Echoing Putnam (1995), who blames TV as the main culprit of the decline in civic engagement in the U.S., some have argued that recent technological advancements, in particular computers and the Internet, have led to young people losing interest and initiative in civic and political participation (Kraut et al., 1998; Wilhelm, 2000). Recent studies, however, have started to challenge such claims, calling attention to the fact that younger generations brought up in the new social context are different from older generations in the way they interact with other members of society and engage in political activities. For example, Livingstone et al. (2005) argue that young people are using the Internet for a wide range of activities that could be considered participatory. They suggest that 'political' participation be broadly defined to include interests not only in party politics and government policies, but also in civic and social activities to do with the environment, animal rights, anti-globalization, gay rights and community activism (Livingstone et al., 2005: 288). Adopting this broader definition, some researchers contend that the informational and communicative functions of the Internet facilitate civic and political participation (e.g., Jennings and Zeitner, 2003; Norris, 2005; Rice and Katz, 2004; Shah et al., 2005; Uslaner, 2004).

This re-conceptualization of political participation in the new media-transformed social context is useful for understanding young people's employment of the Internet for political purposes. Nevertheless, like those earlier attempts at broadening the concept of political participation, it has downplayed nationalistic behavior as an important form of civic engagement and political participation. Such an omission would prevent us from gaining a better understanding of the political self in the Chinese context, where there has been a new surge of nationalism among young people in recent decades.

As Breuilly (1993/1982) has stressed, nationalism is above all political. He notes that nationalism has been explained from various perspectives, such as by reference to ideas, class interest, economic modernization, psychological needs or culture. But, he contends, all these approaches overlook a crucial point, namely that nationalism is above all about politics and politics is about power. The political nature of nationalism has also been highlighted by others (e.g., Hechter, 2000; Zhao, 2004).

Nationalism is political not least in the sense that it is about the negotiation of collective identity. Or as Zhao (2004) notes, situations where nationalism is involved constitute political situations where collective identity is challenged and negotiated, whereby the boundaries between 'we' and 'they' are drawn. The power of nationalism lies in the fact that it 'locates the source of individual identity within a 'people', which is seen as the bearer of sovereignty, the central object of loyalty, and the basis of collective solidarity' (Greenfield, 1992: 3). Furthermore, within authoritarian regimes, such as today's China, nationalism, especially popular nationalism, may represent people's expression of their wishes for democratic participation, which is otherwise unfulfilled.

CHINESE NATIONALISM

Nationalism is a complicated and controversial topic that has attracted much academic attention. The main debate is between primordialism and instrumentalism (Özkirimli, 2000; Zhao, 2004). Primordialism is the belief in the antiquity and naturalness of nations (Özkirimli, 2000; Zhao, 2004). It argues for an essentially unchanging national identity consisting of certain 'givens' of social existence into which one is born, such as shared physical and cultural characteristics, which form the grounds for one's group loyalties (Zhao 2004: 4). Accordingly, nationalist loyalties can be viewed as essentially emotional and 'inherently irrational or extra-rational in the sense that they supposedly violate or transcend considerations of self-interest' (Hardin, 1995: 14). By contrast, instrumentalism argues that the construction of a nation and consequent nationalism is a rational choice of political entrepreneurs in response to specific circumstances. Nationalism, therefore, is explained as a result of essentially self-interested behavior, and nationalist consciousness is seen as a consequence of the historical context in which some interest groups or political forces successfully imagined a political community or national history and persuaded people of artificially shared origins that they were indeed one people: a nation (Zhao, 2004: 4). This approach emphasizes the subjective nature of nationalism, as expressed by Benedict Anderson's (1983/1991) notion of 'imagined communities', Eric Hobsbawm's (1983) 'invented national traditions' and James Wertsch's (2002) 'collective remembering'. It rejects any idea that nationalism is primordial.

Like the term 'identity' (Brubaker and Cooper, 2000), the concept of nationalism serves as both a category of practice and a category of analysis. As a category of analysis, it may seem hard to reconcile instrumentalism and primordialism. However, as a category of practice, it seems that people often adopt both in making sense of their own sentiments and actions. By

the first time in its modern history, with an opportunity to be a real world power (Zheng 1999). It is in this context that the CCP once again revived nationalism as an official ideology by which the government buttressed its legitimacy after the Tiananmen bloodshed, and recruited Chinese people's support in its contention with the actual and perceived anti-China forces and rivals in the world.

According to Hobsbawm (1983: 13), an instrumentalist theorist of nationalism, nationalism, like the other phenomena associated with the nation, such as the state, national symbols and histories, depends on 'the exercises in social engineering which are often deliberate and always innovative'. In the Chinese case, the Party leaders' construction of the 'imagined community' of the Chinese nation-state has involved several innovations. Furthermore, as mentioned previously, in constructing and practicing nationalism, the social engineers have resorted to both a primordialist and instrumentalist approach. First, to prevent secessionist tendencies from ethnic minorities and to reconcile itself with Taiwan, the state invents a pan-Chinese identity in the form of a civic or territorial national identity, which is a supra-ethnic perspective that requires all residents within the country to identify with the exiting state (He and Guo, 2000). It also takes the form of a genealogical identity based on common descent from the Yellow Emperor. Secondly, in the 'collective remembering', which is about inventing national traditions, the leaders have drawn heavily upon pride in China's long history and civilization as well as upon the memory of the country's century of humiliation at the hands of the West and Japan. This process has involved the social engineers' deliberate efforts in reconstructing and reemphasizing traditional Chinese culture and the Party's revolutionary past. Perceptions of present hostility on the part of the West also play a significant role. Moreover, in its construction of the socialist national identity, the social engineers identify China closely with the Communist party (He and Guo, 2000; Zhao, 2005), equating the goals of the state with those of the nation. The purpose is to remove differences within the political community and replacing them with a common, hegemonic order of signs, symbols and values (Zhao, 2004).

The government-sponsored attempts at re-constructing the national community for Chinese people in the reform era was crystallized in the disourse of 'patriotism' (*aiguo zhuyi*) (Xu, 2001; Zhao, 2005; Zheng, 1999). In the 1990s, the Chinese government launched an extensive propaganda campaign to educate Chinese people in patriotism as part of the subjectification regime, as indicated in Chapter 1 of this book. In addition to the earlier-mentioned contents, the regime invested heavily in the so-called 'education in national conditions' (*guoqing jiaoyu*), which Zhao (2005: 135) summarizes in the following:

> The campaign appealed to nationalism in the name of patriotism as a way to ensure the loyalty of a population stewing in domestic

> discontent. At the core was 'education in national conditions' (*guoqing jiaoyu*), which emphasized how China's unique national conditions make it unprepared to adopt Western-style liberal democracy. The current one-party rule, they claimed, would help maintain political stability, a prerequisite for rapid economic development.

The government's construction of nationalism in the name of patriotism has important implications. Patriotism, though closely related to nationalism, should not be confused with nationalism. Zheng (1999) notes that whereas nationalism is based on a shared culture and ethnicity, patriotism requires all those legally entitled to be citizens—irrespective of their ethnicity, national identity and culture—to identify with the exiting state. It thus downgrades ethnicity and culture, and emphasizes loyalty to the existing state (Anderson, 1983/1991). It is in this sense that patriotism constitutes what Benedict Anderson called 'official nationalism', or in the Chinese case, vice versa.

It is noteworthy that although it is predominantly driven and sponsored by the Party state, Chinese nationalism in the reform era is a multifaceted phenomenon, and there are multiple notions of the Chinese nation (Chan, 2005; He and Guo, 2000; Townsend, 1996; Zhao, 2004; Zheng, 1999). This is evidenced by the coexistence of popular nationalism and official nationalism. Popular nationalism, that is, bottom-up national sentiments from the masses, has been flourishing since mid-1990s (Xu, 2001). It has to do with both the 'patriotism education' campaign and the fact that since the 1990s, in its attempt to promote nationalism, the regime has allowed the rise of pluralistic nationalistic voices among the masses despite its strict control over other aspects of political discussions (Xu, 2001; Zheng, 1999). Furthermore, popular nationalism has been reinforced by the Internet, which by facilitating mass communication, has served as a hotbed for such sentiment and action, as mentioned in the introduction of this chapter.

Although often overlapping with and sharing interests with the official nationalism and despite the fact that both want to build a powerful China (Zhao, 2004; Zheng, 1999), popular nationalism constantly challenges the official nationalism by sometimes running against official policies (Liu, 2005; Zhao, 2004; Zheng, 1999). This is evidenced by the attempts among students in China to stage demonstrations to protest against the anti-Chinese riots in Indonesia in 1998. The People's Republic of China government turned down the students' application for demonstrations and punished the students who went ahead with the illegal demonstrations (Zhao, 2002). More importantly, allowing the common people to join with others in forming ideals of equality and dignity in the national environment, popular nationalism has in a sense become the incarnation of democratic process (Xu, 2001). This contradicts the official nationalism which emphasizes the maintenance of the political status quo.

For this reason, Chinese nationalism is a double-edged sword (Zhao, 2004; Zheng, 1999). It can both support and challenge the state structure. On the one hand, the regime can use the nationalistic sentiment from the masses to create a base for political legitimacy and support. On the other hand, without constraints, it could release forces with unexpected consequences, such as popular political participation and national separatism that oppose the regime's goal of nation building. This double-edgedness of Chinese nationalism underlies both the regime's acquiescence during the rise of various nationalist voices and its call for 'rational patriotism' (*lixing aiguo*) (*Xinhua Daily*, April 21, 2008), which postulates that the rational, hence correct, way to show your love to the country is to keep to yourself and do your own duty well. Thus, the official nationalism, or patriotism in this case, brings us back to the regime's emphasis on social and political stability.

'THE NET IS A SPACE FOR RECREATION AND A PLATFORM FOR COMMUNICATION.'

Similar to college students in some other countries (see McMillan and Morrison, 2006), the university students in this study viewed the Internet as an integral part of their lives. Most of them had their personal PC and could access the Internet at the dormitory or at a nearby home. Going online everyday at least for two hours was common among most of my informants. The few who did not own a personal PC visited the Net café at least once a week for a few hours each time. Some mentioned that they could also borrow their classmates' computers in the dormitory.

A shared perception was that 'the Net is a space for recreation and a platform for communication' (*yule de kongjian, jiaoliu de pingtai*), to use their words. Then, how did they personally make use of this 'space' and 'platform'? My informants indicated that for them the Net was mainly a tool for entertainment, although they did occasionally look up things related to their studies. Or as a young man put it, 'Even for us university students, the Net is mainly for recreation, just as for most other Chinese users.' This is in line with the general trend discerned among Chinese netizens. Besides, scanning the front pages of some major websites, such as MSN, Tencent and Sina, was a common practice. Very often they spent much time reading headlines and topics about lifestyles, pop stars and other celebrities as well as national and international news. However, they tended to see such activities as a pastime, as it was not related to their studies. Online communication was predominantly limited to their own circles of friends, acquaintances and families, as in the case of the participants presented in the earlier chapters of this book. This is also in line with findings from studies done in some other countries as mentioned in Chapter 4. Again reminiscent of the other young people in this book, they argued that their communicative activities could be seen as a form of pastime, as they did not usually focus on 'serious' topics. Thus, the boundary between online recreation and communication is also blurred in this case.

Nevertheless, they did recognize the wider communicative functions of the Net. Many of them emphasized that the Net offers, in their words, 'a quick and easy way to exchange information and to connect people together'. This phrase captures the main reasons why many people have expected the Internet to serve as an effective means for political participation. However, the extent to which this potential is exploited depends on the attitudes of citizens and governments towards the Internet, which can be quite different in different societies, particularly in terms of using the Internet for political participation (Chase and Mulvenon, 2002; Taubman, 1998). These young people's perceptions and experiences of the Internet as a political tool are specific to the socio-cultural context of today's China, or the technology's receiving context (Zhou, 2006: 232), which is defined as 'the multi-layered, constantly changing historical environment in which the technology is received, adopted, controlled, and applied'.

'EXCUSE ME. WHAT IS THE CENSORSHIP ABOUT, PLEASE?'

In airing their views about the Internet's impact upon Chinese society, my informants distinguished between the 'personal' and the 'national' Internet. They saw the Net as mainly a space for recreation and secondarily, a platform for communication for users—often with the overtone that it was not essential, albeit important, for individuals. However, they believed that the Net is playing an irreplaceable role in China's national development. Typical remarks are such as: 'It is crucial for China's development', 'It allows China to enter the world and the world to enter China.' and 'It is the Information Age now. Without the Internet and other high technologies, it would be impossible for China to compete with other countries.' Such views reflect the official discourse discussed in Chapter 2 which frames the Internet as the locomotive for China's economic development and overall modernization project. However, promoting democratic participation with the technology has not been the intention of the Chinese government.

The young people's agreement with the official discourse lends support to the observation that China's netizens, who are mostly urban youth and who belong to the new middle class, generally show affiliation with the government, as they are the beneficiary of the reform and opening up (Damm, 2007). They identify with the government in its efforts to establish the Internet as part of a modern economy. My informants showed great pride in the achievement China has made in the past years in information technology, which according to them, like the success in launching the rockets into the space, is indicative of the country's growing power. In short, for these young people, as for the Chinese society in general, the Internet is largely a means for national economic modernization and personal recreation, rather than for political modernization.

When asked 'How has the Internet, as a more open medium of information and communication, influenced your political views, civic and democratic participation?', a predominant reply was 'No effect on me.' In reporting about their online communication, my informants usually indicated that they seldom, if ever, participated in online discussion forums (such as bulletin board systems or BBSs). The few exceptions show that if one regularly visited a certain online discussion forum, it was mainly for the purpose of discussing game-related topics and/or maintaining 'real life' hobbies. For example, a young man mentioned that the only online forum he had participated in was the one focusing on an online car-driving contest, because he was fascinated with the game and had even won a prize from it. For another young man, who was interested in fishing, an online discussion forum on fishing proved helpful in maintaining his hobby.

In talking about their experiences of the Net, no one mentioned the control of by the government. When asked 'What do you think of China's Internet censorship and how does it affect your use of the Net?', the majority of the informants said that they had not heard about the censorship. Many of them asked me instead, 'Excuse me. What is the censorship about, please?'. Among the few who showed some knowledge about this system, only one said, 'It appears a bit authoritarian.' The others insisted that it did not affect their use of the Net at all, because 'The things we do online, such as playing games, are not related to censorship', they claimed. Another informant explained, 'The censorship is against pornography. We just play games. So that is not relevant to us.' Moreover, according to some of them, 'the Internet is already a platform for people to say whatever they like to say' (*changsuo yuyan*), implying that they do not feel there is anything lacking or limiting. But not everyone likes this 'freedom'. Rather than wishing for a freer Internet, some of them expressed a strong wish for a 'real name' system, not only to better protect people from harm due to anonymity but also to exert a stronger control generally over what people can say in cyberspace. Whereas it may be a good intention that people be responsible for what they say in cyberspace, such views legitimize the governmental control. It is ironic that what the young people perceived as problematic in relation to the Internet seems to be quite different from the concerns that Western observers have about the governmental control of the Net.

'I AM NOT VERY MUCH CONCERNED ABOUT POLITICS.'

The previously outlined understanding of the Internet as a political tool captures only part of the story of why the young people had rarely employed the Internet for political participation (as narrowly understood). This had to do not only with their perceptions of the Internet itself. More importantly, it was related to their political self in the offline context, especially how they perceived political participation.

Reflecting upon the Internet and political participation, most of the informants indicated that they were not interested in politics. Or as some put it, 'I am not concerned about politics.' To understand this 'indifference to politics', it may be useful to investigate how they understand 'political participation'. In response to this question, most of them replied, 'I do not know.', 'Have no idea', 'Have no such a concept in my mind.' and so on, although a few did name one or two items, such as voting, joining the party and participating in a rally and so on. A probe of whether they had participated in politics or planned to do so in the future yielded a predominant reply: 'No. I don't think so.' Nevertheless, from the reasons they offered for this negative reply, one can see that they did have their own understandings of what political participation means in the Chinese context, as is reflected in the following remarks:

> It is no use even though you make some suggestions or criticism. They will not adopt it. So it is better not to bother about that. People would say you are interfering into things that do not concern you (*duo guan xianshi*). Or you are not observing the 'proper' norm (*bu shou benfen*).

> Chinese people today are all relatively selfish (*zisi*). For Chinese, it is not necessary to do a thing if it is not beneficial to yourself. Nowadays, people are pragmatic.

> Politics is not practical. It is very hypocritical and full of corruption. I feel it is far away from me.

> I do not vote. Only party members vote. As long as I live well, I do not care who the President is.

> I do not want to be influenced by the *Falun Gong* people. I trust the Chinese Communist Party much more than the *Falun Gong* people.

One can see that political participation from these Chinese young people' point of view is a rather narrow concept. It is largely limited to engagement with the government, especially anti-government activities, such as criticizing the government, making suggestions to the government and attempts at subverting the CCP's rule. This seems to be much in agreement with the opinions of the elderly Internet users in a study conducted by Xie and Jaeger (2008) and with the narrow definitions of the concept that limit political participation mainly to engagement with the government. It is understandable that people just choose not to participate given that even if one cares to take such actions, one will not be able to make a difference, and worse still, one may cause trouble for oneself. Or as a young man emphasized, 'participating is one thing, participating effectively is quite another. When one cannot make a difference, why bother?'. Moreover, if politics is seen as a complex and 'dirty' game, as it is 'hypocritical' and 'corrupted', not getting involved may represent one's declaration not to mix with the 'bad'. In addition, in the case of present-day China, where the economic success has generated newly emergent confidence among Chinese people in the party leadership for a bright collective future (Zhang, 2004), engaging with the government may be viewed as disturbing the hard-earned social stability both by the government and by society at large. The wish to keep the status quo is reflected in another shared understanding that, as a young man put it, 'It takes time to develop. The leaders are doing a good job. China has its own characteristics. China cannot copy the West, and vice versa.'

However, the 'political indifference' expressed by the young people does not necessarily mean that they totally neglect politics. In declaring a break away from the governmental politics, their choice is already political in effect. Furthermore, as I shall show in the following, their enthusiasm about and active participation in nationalism on the Internet adds quite another dimension to their political self.

THE ANGRY NATIONALISTS GOING ONLINE

Though they professed lack of interest in what they perceive as 'politics', my informants nonetheless did employ the Internet for vehement nationalistic expression. Talking about the incidents related to the Beijing Olympics, each and every one of the informants said, 'I was enraged and disgusted!' Many of them reported that they signed some online petitions calling on Chinese all over the world to support the Olympics against the 'separatists'. Nearly all of them took part in similar activities organized by their university and/or by their online communities. All indicated that they were alerted to the development of these incidents and kept a close watch on relevant information via the Internet and TV. Many of them were among the millions of Chinese netizens who attached a 'red heart China' or 'I love China' icon to their online profiles. Some changed their online names into phrases or expressions to signify their love and support for China, such as 'I Love China' (*wo ai zhonghua*) and 'Powerful China' (*qiangda zhongguo*). A few of them reported having posted comments on BBSs to let out their anger and to encourage Chinese people to unite and make the country powerful. For example, a young woman said:

> I was so angry. Our national dignity was insulted by CNN, Dalai and the West. The West represented by the US is always afraid of China

> becoming strong. They would grasp every chance to disturb us. . . . We Chinese people have suffered too much in the past. It is high time that China became powerful. I sent out many postings in the BBS to cheer up the Chinese people.

Another shared experience among the informants was that expressing their anger over the 'separatists' and the West constituted the main content of their online communication within their own circles during those days. They were busy passing on messages to organize people to sign protests, petitions and boycotts. They recalled that one frequently got and sent out messages like 'if you are a Chinese, you ought to read this, and/or sign this' and so on. Some of the informants showed a strong belief in Internet-organized 'political consumerism' (Stolle, Hooghe and Micheletti, 2005) in the form of boycott as an effective way to teach the anti-China individuals and organizations a good lesson. A young woman who had signed up for boycotting both Dior and Carrefour, told the story about how the American celebrity Sharon Stone was heavily condemned by Chinese netizens when she indicated that the May 12 earthquake was karma for Chinese people's 'wrong deeds'. This informant told me many of her female friends and she herself had stopped using Dior, which has Stone as a spokesperson, in response to an online petition by a female Internet user. 'We cannot affect her power directly, the only thing we can do is to make her lose economically.' she added.

Through such actions targeted at the 'anti-China' forces, the young people experienced the Net as a powerful tool to unite people together for a common goal. Some of my informants mentioned that they had been experiencing an unprecedented sense of belonging and collective power those days. And this was another topic that they frequently took up in their online communication. One of them told me:

> Chinese people have not been so united (*tuanjie*) for many years. People have been divided since the reform and opening up started. All have been bent on their private interests in the competitive market economy . . . We should thank the anti-China forces for making us so united. The more they are against us, the more we show solitarity. This has greatly encouraged us Chinese. They should study Chinese history before they claim that Tibet is a country. This is ridiculous!

I came across similar opinions on many other Chinese websites both inside and outside of the country, where there were many postings in praise of the re-found solidarity among Chinese people.

It is, thus, not hard to discern the delight and pride the young people had found in their 're-gained' collective identity as Chinese. What has made them so nationalistic and yearn so much for assertion of a collective Chinese identity? A good start for understanding this question is to explore how they understand nationalism or the Chinese term '*aiguo zhuyi*' (patriotism), which they used to refer to one's love for his/her country/nation (*guojia*).

'AIGUO ZHUYI IS OF GREAT SIGNIFICANCE.'

My informants were articulate in talking about *aiguo zhuyi*. All emphasized that *aiguo zhuyi* is of great significance. They provided explanations that are at the same time primordialist and instrumentalist. They strongly shared an understanding that their love for the country is natural, or as some of them argued, 'I love China because I am a Chinese. It is in the nature (*tianxing*).' With this primordialist assumption, they argued ardently that to love one's country is a most basic quality (*suzhi*) as a human being and that this moral-normative dimension of '*aiguo zhuyi*' is only comparable to one's devotion to one's family. Indeed, they saw the country as a family and people who live in it as its members, echoing the official concept of a pan-Chinese nation, with 56 ethnic groups, who are brothers and sisters within a big family. It is, thus, they argued, only natural that one love and be loyal to the family, even if the family is not perfect. Furthermore, just as its members' interests are closely tied to the family's flourishing, one's own fate will definitely be affected by the nation's development. This kind of loyalty to one's country constitutes what Fong (2004b) refers to as 'filial nationalism' found among Chinese young people.

Some carried this primordialist explanation of their *aiguo zhuyi* even further, arguing that such a kinship bond among people is a unique characteristic of the Chinese people, which one should be proud of. A young man made the following remark:

> It is Chinese people's nature to unite themselves against outsiders once there is a common enemy although there may be very bad conflicts among themselves. This is in our blood. That is what Chinese people are like, quite different from Westerners.

For these Chinese youth, then, the nation is not constructed for instrumental purposes, nor is it felt to be the 'imagined' community that Anderson (1983/1991) wrote about. Rather, it is real and tangible, embedded in common roots and kinship bonds.

However, at the same time as they emphasized this natural givenness of their love and loyalty for China, some of them also alluded to the acquired aspect of their *aiguozhuyi,* by explaining their *aiguo zhuyi* from a socialization point of view, pointing to the 'patriotic education' campaign since 1990. For example, a young man said, 'This has to do with the education we have been through. From the beginning of our schooling, we have been exposed to *aiguo zhuyi* education. It is hard not to be influenced'. Furthermore, they claimed that *aiguo zhuyi* should be the most basic quality (*suzhi*) of a citizen too, not only that of a human being. In other words, it is a consciousness which can and should be cultivated in every member of the nation-state. With this acquired consciousness, many of them uttered a declaration-like remark: 'Of course, I love my country. As a Chinese citizen, I will definitely not cower when my country needs me. This is the most basic.' According to my informants, such loyalty to the nation-state, and obeying the law and maintaining social order are the most important signs of a qualified Chinese citizen.

Thus, although scholars on nationalism tend to see the primordialist and instrumentalist approaches as contradictory, it seems that these Chinese young people used them complementarily. Such understandings echo the aforementioned social engineers' construction of patriotism in the post-Mao era. Nevertheless, it would be misleading to believe that the young people's notion of nationalism is totally identical with the Party leaders' discourse of patriotism. As we shall see in the following, their perception of *aiguo zhuyi*, even though they used the same term as the leaders do, goes beyond that of the official nationalistic discourse.

'ONE MUST DISTINGUISH THE NATION (GUOJIA) FROM THE PARTY (ZHENGDANG).'

As mentioned earlier on, the Chinese state or official nationalism identifies China closely with the Communist party and it equates the goals of the state with those of the nation. In contrast, although their understanding of *aiguo zhuyi* reflects very much the patriotic socialization they have been exposed to (such as their antagonism to the 'anti-China' forces and 'separatists', and their wish for a strong China), many of the young people in this study deliberately emphasized that one must distinguish between the Party (*zhengdang*) and the Nation (*guojia*) in talking about *aiguo zhuyi*. A young woman expressed a shared view:

> Loving the country and loving the Party are two different things. I'd like to give just one illustration. The country is over 5,000 years old now. However, the current government is only several decades old.

Another student put it 'The nation is higher and greater than the Party government. They may not represent the same things, although these two do not need to contradict each other.' And exactly because what the nation represents transcends what the Party stands for, as one student put it, 'The country's honor (*rongru*) is considered higher than the interests of the Party.' For the same reason, *aiguo zhuyi*, according to my informants, should not only be outward directed, but also inward directed. This means, some of them explained, that actions may be taken against the Party leaders if they are harming national interests the same token, although as a category of practice, Chinese nationalism resorts to both the primordialist and instrumentalist approaches, as a category of analysis it seems best explained from an instrumentalist perspective, as I shall show in the following.

The origins of Chinese nationalism remain disputable. Nevertheless, there seems to be a prevailing consensus among scholars that before the late nineteenth century, when China was still an empire, nationalism did not exist, because the Confucian image of China was that of a culturally defined community rather than an ethnically/politically defined nation-state (Pye, 1990; Zhao, 2004, 2005; Zheng, 1999). Nationalism emerged only after China's disastrous defeat by British troops in the 1840–1842 Opium War (Zhao, 2004; Zheng, 1999), which forced Chinese to change their perceptions of state power and world order. There was a growing awareness among the Chinese elite that China ought to be seen as a nation situated within the 'social Darwinist' capitalist world system rather than a self-complacent Confucian center of civilization (Chang, 1987; Harrison, 2000; Fong, 2004; Tang, 1996; Wallerstein, 1998). It was a shared understanding among the political and academic elite that only nationalism could save China by preventing it from falling victim to conquest and colonization. The concept of nationalism requires that Chinese see themselves as belonging to 'China', rather than to tribes and regional and lineage-based groups (Fong, 2004).

It was against this background that China's modernization and nationalism began, the goal of both being a strong state and surviving as a nation. Throughout modern Chinese history, building a strong nation-state was a consistent theme of Chinese nationalism, leading to a 'strong state complex' among social and political groups (Zhao, 2004; Zheng, 1999).

Although Mao has never rejected this 'strong state complex' (Zheng, 1999), during the Cold War his attempt to remove China from all world systems resulted in nationalistic awareness being largely replaced by a resistance to integrating into the capitalist world system (Fong, 2004b). With China's reform and opening up at the end of the 1970s, especially since the 1990s, China has seen an outburst of nationalism. As Zheng (1999) argues, many factors at home and abroad have contributed to the rise of the new nationalism in post-Mao China. It is not only a result of China's rapid economic growth and its growing power. More importantly, it is a response to the 'Chinese problems' (both internal and external) that post-Mao China has encountered, such as the crises in the faith in Marxism and Maoism, the decline of central power, the weakening of national identity, and actual or perceived external threats (Zheng, 1999). Stripped of communist ideology as its guiding light and lacking the procedural legitimacy of democratically elected governments, the Chinese Communist Party (CCP) is increasingly dependent upon its nationalist credentials to rule and hold the society together (Gries, 2005; Zhao, 2004, Zheng, 1999). Moreover, the end of the Cold War and the collapse of the former Soviet Union provided China, for through, for instance, corruption or stupid foreign policy. This echoes the claims made by participants in the 1989 pro-democratic protest, who claimed that their action was patriotic. It follows that now that these same young people showed little interest in engaging with the government, it may mean that they are basically content with the current government. However, it is worth noting that the effect of the political control may have also played an important role in shaping the Chinese political habitus.

The university students' nationalist enthusiasm may have important bearings upon their collective identity, China's relationship with other countries and not least, the individuals' relationship with the current government. In this sense, their nationalistic expression, be it online or offline, is political in nature. Nevertheless, the distinction between the nation and the Party has led them to contend that *aiguo zhuyi* has little to do with politics. Some argued that *aiguo zhuyi* is morally superior to political participation. Or a young woman put it, '*Aiguo zhuyi* is a noble and commendable thing whereas participating in politics may not be so.' This agrees with their perception of political participation, which to them, mainly means engaging with the government and which often has a series of negative connotations in the Chinese context.

Despite their view that *aiguo zhuyi* is not political, they did see online nationalism as a demonstration of democracy, because, they claimed, it is participatory in nature and everyone (with Internet access) has an equal chance to say something and have a voice. This collective power through mass equal participation, according to my informants, lay behind CNN's final apology to China. It was the power of Chinese netizens, who were connected by the Internet, that had made the difference, my informants insisted. This echoes an article posted in the Strong Nation Forum (*Qiangguo Luntan*) on April 25, 2008 entitled '*Not Afraid of the Mainstream Media, but Afraid of Chinese Netizens*'. The author makes the point that the Internet has had such power exactly because it is a mass-participatory medium in sharp contrast with the other forms of media which are tightly controlled by the government. Thus, according to the author, the apology from CNN was a result of the confrontation between Chinese netizens, rather than China's official media (or government) and the West. The author calls on the government and society at large to reflect upon the role mass participation can play in defending the country's interests and supporting the government. This role, the author argues, should form the grounds for the government to treat the netizens, who represent the people, well.

Taken together, the young people's creative and active use of the Internet for nationalistic actions was not only an expression of their patriotic sentiments, but also a sign of their wish to exert influence upon national affairs. Thus, the Chinese case has added another dimension to the political character of nationalism.

CONCLUSION

The Internet holds significant potential to affect current understandings and activities of political participation. Even in China, where the Internet is under strict governmental control, it has facilitated a public space, where people (with access) can engage publicly with certain social and political issues.

It has been widely observed that Chinese netizens' present use of the Net is overwhelmingly apolitical, with a focus on entertainment, lifestyles and private issues. And yet, this depiction neglects another striking theme in Chinese cyberspace, namely nationalism (Qiu, 2003; Wu, 2007), which constitutes the 'most viable political discourse among Chinese netizens that is both routine in everyday discussion and of fundamental importance to the dynamics of social movements in China's cyberspace' (Qiu, 2003: 15). This phenomenon was especially striking in the wake of the incidents surrounding the 2008 Beijing Olympics.

The coexistence of the two contrasting themes in cyberspace has implications for an understanding of political participation among youth, hence their negotiation of a Chinese (political) self within the specific socio-cultural context of China's dual modernity. What makes Chinese netizens indifferent to other forms of political involvement while they are enthusiastic about nationalism? In this chapter, I have explored this question by investigating how a number of university students, who belong to China's future social-political elite, perceived and experienced the Internet as a political tool, which was closely related to how they understood political participation and nationalism in the Chinese context. This is done by locating their narratives within a re-conceptualization of political participation which takes into account nationalism as an important form of political expression. A re-conceptualization of nationalism as indeed 'political' is useful for gaining such an understanding given that previous studies on contemporary Chinese nationalism and on political participation have treated the two as largely isolated from each other.

In line with the general trend observed among Chinese users, my informants were using the Internet mainly for entertainment and communication with friends, acquaintances and families. Their 'apolitical' online activities need to be understood against the socio-cultural context in which both the Internet and users' lives are embedded. In an authoritarian society like China, the government does not intend to use the Net for promoting democratic expression and participation, but for China's economic development and the overall modernization project. The censorship, which has been designed to curb the liberating effect of the mass communicative nature of the Internet, represents attempts to discipline the Chinese mind along this line. This has coincided with a range of other factors, which all work together to shape youth's political attitudes. First, this seems to reflect the post-1989 'depoliticalization' or 'political pragmatism' widely found among Chinese young people (Damm, 2007; Rosen, 2009). The lesson from the Tiananmen Square event, along with the flourishing consumerism accompanying the reform era, has taught Chinese people to care about their own well-being and comfort. Secondly, the widely perceived 'political inefficacy' and a tendency to take comfort in conformity (Farquhar and Zhang, 2005; Xie and Jaeger, 2008) reflect the Chinese habitus nurtured both in the long Confucian tradition and the Communist regime. They keep conveying the message that one should 'keep to oneself' (*anfen shouji*), rather than disrupting the 'harmony' of society and running the risk of causing trouble for oneself. The tendency to conform to the mainstream and maintain the status quo has also been encouraged by the discourse of a 'harmonious society' as mentioned in Chapter 1 and by the flourishing economy in the reform era that lends legitimacy to the Party leadership. This tendency may be especially evident among university students who are the beneficiary of the reform and opening up, and who in many cases may see themselves as prospective members of the emerging elite. These factors, together with the governmental control of the Internet infrastructure, access, and online content account for the young people's 'indifference' to political participation in terms of a narrow definition of such participation.

The political self takes on quite another dimension with regard to use of the Internet for nationalistic expression. Although they have rarely used the Internet to engage with the government, they reported having taken an active part in the online nationalist activities concerning the incidents surrounding the 2008 Olympics. This has to do with their perception of nationalism, and of what it means to be Chinese within the current nation-state. Their perception of nationalism, *aiguo zhuyi*, to a great extent echoes the patriotic education initiated by the government in its attempt to reconstruct China as a national community in the post-Mao era. Like the official nationalism, their account of their strong *aiguo zhuyi* was from both a primordialist and instrumentalist perspective. In so doing, their emphasis was on loyalty and duty as a Chinese citizen and as a member of the nation as a big family. This loyalty towards one's nation-state, together with obeying the law and keeping to oneself, which according to my informants is another important indication of civic awareness, seem to show that their understanding of 'citizenship' differs from how the term is understood in democratic societies. Civic awareness in democratic societies means people's awareness about and participation in social and political affairs, sometimes involving defense of the individual's interests against possible governmental encroachment. Although in both the Chinese case and in the West, patriotism is expected to play a role in sustaining a higher ideal of citizenship (Stapleton, 2005), the Chinese understanding of citizenship, with its emphasis upon 'obedience' and loyalty, hence on duties rather than rights, reminds one of the 'subject' rather than the 'citizen', thus begging the question whether or not the Chinese young people, even the educated elite, have assumed the role of a citizen in its Western sense of the word.

Nevertheless, although these young people's perceptions of nationalism are very much aligned with the state agenda, it would be misleading to believe that the two are identical. With their insistence on the distinction between the nation and the Party/government, their understanding that *aiguo zhuyi* should be inward directed as well as outward directed, and their celebration of the Internet as a platform for mass democratic participation, their *aiguo zhuyi* meanwhile reflects the popular nationalism, which may also be construed as showing a liberal tendency (Zhao, 2005), coexisting with the official nationalism in today's China.

In this sense it can be argued that nationalism in China as reflected by the case of the university students in this study represents a special form of popular participation under the post-Mao condition. Thus, it may serve as an enclave of democratic public life (Xu, 2001). Furthermore, vehement nationalism, like other efforts in community building, represents a search for unity and collective identity within a context of social fragmentation and individualization. The Internet provides a public space, rarely existent offline, for such expressions. For these reasons, young people's nationalism constitutes an important extension of the political self in the Chinese context. This means that to gain a better understanding of the Chinese political self in the Internet age, we need to locate users' seemingly contradictory online behaviors within a continuum rather than treating them as isolated phenomena, especially given the kind of 'dualism' observed among Chinese youth, a theme that I shall return to in the concluding chapter.

Conclusion
Modernity, the Internet and the Self

Spectacular transformation has occurred in Chinese society since the late 1970s, when China started its reform and opening up with its pronounced modernization project. The popularization of new information and communication technologies, especially the Internet, has further complicated this social drama. It is against this backdrop that the younger generation's life journey is unfolding. This book has sought to show that investigating the interaction between Chinese youth and the Internet, both of which are strong driving forces for social change (at least potentially so) is a useful way to enhance our understanding of Chinese society.

The investigation was organized around two general themes, namely Chinese young people's perceptions and experiences of the Internet and their online self-representation. I have explored how urban youth in a medium-sized city in North China were actually taking up the Internet and making it fit into their everyday lifeworlds and presented two case studies respectively on Chinese youth's online community building and their political participation in the Internet age. It is a central thesis of the book that young people's experiences and perceptions of the Internet and their online identity work are closely related to their everyday life actualities embedded in the larger socio-cultural context of today's China, which is undergoing dramatic transformation due to a host of local and global forces.

With the limited number of cases analyzed and the purposive sampling, this volume does not claim to be a definitive record of Chinese youth culture and, related to it, youth's Internet-related experiences, both of which are complex, fluid and constantly changing. Rather, it provides a few snapshots that serve as contextualized examples of it. In this concluding chapter, I will draw together the major threads that have emerged from the preceding chapters, focusing on three major interrelated themes.

WIRED FOR FUN: THE 'DOUBLE-FACED' YOUTH AND THE INTERNET

The major findings from this study support the argument that there is no such a thing as a homogeneous 'Net generation' across the globe, just as there is no such a thing as the 'universal' Internet. Different users can employ the Internet in different ways depending on their social-biographical situations and relevance systems. That is, both users' Internet-related experiences and perceptions, and their online self-representation reflect their everyday life actualities embedded in the larger 'glocal' context of their societies.

Viewed from such a perspective, it is no surprise that Chinese youth seem to orient themselves to the new medium in ways that seem to be strikingly different from their Western counterparts, despite similarities. Two trends in the Chinese case are worth revisiting in comparison with some Western societies. First, the vast majority of my informants employed the Internet mainly or purely for entertainment. They simply defined the Internet as 'entertainment'. Whereas the Internet has acquired high relevance in this realm of their everyday lifeworlds, it has gained much less importance as a device for information and learning purposes. Most strikingly, going online was viewed largely as an antithesis to study (and by extension, work and other 'proper' businesses of life). Such a recreational orientation towards the Internet seems to reflect the global youth Internet culture (Buckingham, 2003; Livingstone, 2003; Sandvig, 2006; Sefton-Green, 2004; Seiter, 2005), where young people seem far more fascinated with the Net as a leisure device than an information and educational device. However, Chinese youth seem to employ the Net for entertainment to a greater extent than their counterparts in some Western societies. For instance, research done in the U.K. and Norway reveals that young people regularly use the Net for information seeking related to their school work as well as for recreational and social purposes (Liu, 2010; Livingstone and Bober, 2005), with a majority claiming that the Internet is an effective helper with their academic performance. In a comparative study I conducted (Liu, 2010) on Chinese and Norwegian high school students concerning their relationships with the Internet, the Norwegian participants found it hard to assign greater relevance of the Net to entertainment than to school-related uses, although, like youth in other societies, they were also fascinated with the Net's affordances for recreation and communication. To be sure, they also needed to negotiate between the Net's relevance as a leisure device, which is more intrinsic to their interest, and the Net as a learning and information device, which can be seen as the imposed relevance of the Net (Liu, 2010). Nonetheless, most of them seemed to have a more balanced relationship, whereby they employ the Net equally for the various purposes—recreational, social and instrumental—compared to their Chinese counterparts, who only, or at least mainly, used the Net for entertainment.

Secondly, my informants tended to see the Internet as 'another' world of its own right paralleling this world. Most of them just defined the Net as the 'virtual world' and they emphasized that it was exactly the 'virtual' nature of the Net, or the differences and contrasts between 'that world' and 'this world' that fascinated them. In that 'other' world, via what they lumped as online entertainment, they engaged in various activities that were related to their inner self, ranging from fun seeking to displaying individuality, relaxing, escaping from 'this world', letting out bad moods and experimenting with identity. Thus, their relationship with the new medium was mainly 'expressive' rather than 'instrumental', invoking a user-Internet relationship in which the Net is seen as a place and a way of being rather than a tool (Markham, 1998). The predominantly 'expressive' orientation towards the Net indicates 'an increasing investment of the self in the Internet where cyberspace becomes an arena for action and interaction as important as user's offline life' (Bakardjieva, 2005:104). This forms a striking contrast with their Norwegian counterparts, who saw the Internet as an extension of 'this world' rather than 'another' world (Liu, 2010). Such a difference is in line with the findings from the *Young Digital Maven* study, a large-scale survey mentioned in Chapter 2, which shows that compared with their counterparts in the U.S., Chinese youth display a significantly greater drive for online creativity and experimentation with self-expression and identity.

The way in which the majority of my informants oriented towards the Net reflects the general Chinese Internet culture discussed in Chapter 2, which has been nurtured by the predominantly consumption-oriented new media culture and which features entertainment, consumerism and lifestyles, reflective of the 'consumerist postmodern' society of post-Mao China. However, this pattern needs also to be viewed against these young people's social-biographical situations and relevance systems, which are reminiscent of the general social-biographical characteristics of Chinese urban youth. As I show in Chapter 3, reflective of the material and cultural conditions under 'socialism with Chinese characteristics', with its inherent hybridity, non-synchronicity and contradictions, Chinese urban youth's identity construction seems to revolve around an interrelated set of apparent paradoxes, or double-facetedness, which together constitute a complex and multi-faceted collective character. This observation concurs with Chinese youth's self-representation, as reflected in the post-80ers' collective narrative online with a focus on 'dual qualities'. They are simultaneously materialistic and idealistic, instrumental and expressive, internationalist and nationalist, global and local, apolitical but nationalistic, modern and traditional, blessed with material prosperity and yet under great pressure to strive for greater material wealth, and exposed to an unprecedented range of choices and yet lacking a sense of security. Most strikingly, material welfare and great pressure coexist in their everyday lifeworlds and they come across as simultaneously embracing the two largely contradictory forms of individualism: expressive individualism and economic/utilitarian individualism, as described by Bellah et al. (2008).

Chinese youth's predominantly recreational orientation towards the Net, especially in the case of some of the young adults who could afford 'hanging out on the Net' for a considerable time daily, mainly for recreation, is a manifestation of the well-being enjoyed by Chinese only-children and it reflects their status as urbanites and members of China's emerging middle class. For them, 'emerging adulthood' is becoming a reality and leisure consumption is increasingly integrated with high-tech as in wealthier societies. Like the consumption of other commercial goods, ranging from Western fast food to the family car, 'surfing' the Net constitutes 'cool' consumer practices for urban youth. Their access to the Internet and other new media alone is indicative of the general material and cultural advantages that their rural counterparts can only be envious of. However, underneath the seemingly hedonistic approach to the Net found among most of the participants was a self under pressure, frustration and boredom. My informants readily associated their entertainment-oriented relationship with the new medium with being unable to bear the too great pressure in the 'real' world. They frequently referred to being the only-child, being young, being Chinese, living in a relatively small city, lacking alternative leisure space, lacking development opportunities and lacking space for individual expression as the everyday realities that have framed their 'choice' about how to relate to the Net. In telling their 'I-and-the-Internet' stories, the participants drew different aspects of these situational characteristics into the foreground when assigning relevance and meaning of the Internet to their individual lives. They claimed that because of the pressure, frustration and boredom, one needed to resort to the 'other' world of cyberspace for relaxation and for an escape. The perception of overwhelming pressure is also a strong theme emerging from the post-80ers' collective narrative, as seen in Chapter 7.

The constraints upon their lives due to the earlier-mentioned factors have led the young people to compare China with the West. This tendency of discussing things in China by referring to the West, which they tended to perceive as the ideal 'other' of China, might have to do with their knowing I returned to China from a Western country to conduct this study. However, I think it also reflects a general tendency among Chinese young people to compare China with the Western developed societies as a way to voice their discontent with the ways of life in their local context. Such comparisons throw in sharper relief the contrast between the two, thus causing more pressure and frustration for the young people who find themselves trapped in a less satisfying situation. Chinese urban youth's admiration for the West is not limited to the latter's material condition, which is the focus in Fong's (2004b) study of teenagers in Dailian, but also in terms of its cultural civilization, not least individual freedom, which constitutes a striking theme in this study. This tendency reflects the dilemma of being both 'local and global' at the same time, a situation in which many young people around the world find themselves in this age of globalization, whereby the image of the 'ideal' other represented to them via the mass media and other globalization processes shape their 'glocal' imagination. Furthermore, given the relatively disadvantaged condition of a less developed city compared with the big and more developed ones, one can also sense a discontent among some of my informants, albeit not very strongly uttered, with the limitedness of living in a 'small' city. As such, it seems that compared with young people in big cities, youth in more remote urban areas are under double pressure to identify with both the big cities within the country and the Western wealthy societies. It is interesting to note that sometimes Chinese young people's romanticized version of Western life may prove to be misleading given that young people in Western societies may also find their lives highly structured and constrained in terms of both physical and social spaces (e.g., Boyd, 2008; Sibley, 1995; Skelton, 2000; Skelton and Valentine, 1998).

The widely shared sense of pressure, frustration and boredom is also related to another trait of Chinese urban youth: although eager for individual self-expression, they are, nonetheless, also dismal pragmatists bent on the goal of the 'middle-class dream' based on material accomplishment as measured by certain pre-defined indicators. The latter contradicts their pursuit of the 'cool' with its emphasis on individual differences. One can probably find such contrasting elements in young people in other societies as well. However, it is the forceful intensity with which Chinese youth are displaying both themes that is intriguing. This forms a contrast with the individualization process in the U.S., where the two forms of individualism have manifested themselves as two stages of one process, according to Bellah et al. (2008). In my view, China's younger generation's embracing expressive individualism and economic/utilitarian individualism at the same time and with equally intense alacrity constitutes an important feature of the individualization process in post-Mao China. Such double-facetedness has implications for young people's identity work, as reflected by the participants in this study. The sense of pressure, frustration and boredom can be an effect of the utilitarian demands on the expressive self. For Chinese urban youth, negotiating a modern individual identity is then a matter of how to balance the two forms of individualism. The eagerness for individual expression only becomes the more tempting when one is exhausted by the pressure incurred by utilitarian individualism. But it is also possible that the more one is individualistically expressive to begin with, the more keenly the pressure, frustration and boredom will be felt.

Chinese urban youth's eagerness for expressive individuality has to do with the diversification of value systems in the post-Mao era. As China's 'wired generation', which draws its icons, styles, images and values mainly from the Western consumer cultural production and entertainment industry, Chinese youth are getting more like their Western counterparts in value orientation and outlooks, eagerly pursuing the '*ku*' ('cool'). The same process of becoming 'modern' and 'global' has also changed their notion of what constitutes 'the good life', which requires considerable individual efforts to achieve for most Chinese. The temptation of the Western middle-class lifestyle is compounded by high parental expectations of the only-child. As Tsui and Rich (2002) point out, with a Confucian tradition that stresses the importance of education, the pressure for the only-child to achieve has been intensified by an increasingly urgent sense that the family is the most reliable 'welfare agency' for its members. Parents wish to 'transfer to their child whatever cultural capital they have and purchase more if possible, and thus to maintain or improve their family's social position in a more unequal and competitive society' (Stockman, 2000: 112). Under such circumstance, the Chinese only-child, whose life is from the very beginning a project for parents to work on (Liu, 2006a), can hardly opt to follow his or her own intrinsic interest. The life trajectory designed for them by their parents, together with the stifling educational system and the strong pressure to make material accomplishments, tends to leave little space for individual expression. This is contradictory to their urge for individual freedom and differences.

The tension to which Chinese urban youth are exposed largely shapes both their identity construction in general and their choice and meaning making in relation to the Internet in particular. The Internet has not only become a venue for articulating their expressive individualism. It also is a means to cope with the pressure due to the tension between the two forms of individualism. This explains why for Chinese urban youth, the Internet serves as another world, which is depicted as more beautiful, freer, easier and simpler than 'this world'. The Net provides compensation for the 'lacks' in the 'real' life and a temporary escape. The same accounts for the popularity of the Internet café for many urban youth, who actively help turn it into a 'heterotopian third place', which serves as an oasis among the overwhelming demands of modern life. 'Playing' online, thus, not merely reflects young people's tendency to indulge themselves in fun seeking. It also represents a form of resistance to constraints.

Given the potential of the Internet for information and learning, one might expect a good fit between Chinese urban youth's achievement goals and using the Net. However, the reality seems largely to be the opposite. This has much to do with how success is defined and evaluated in the reform era. Despite the discourse of free choice, young people's life chance remains highly structured by the educational system, especially the National College Entrance Exam. The exam mainly takes the form of pen-and-paper tests based on textbook knowledge, allowing little room for exploration beyond the textbooks. We have seen in this book how such an education system underlies urban parents' worry about the Internet's distracting their children from schooling and from other activities necessary for success and why there is a widely shared perception that the Internet is harmful for students. Again, such a perception seems to stand in sharp contrast with some Western societies. For example, based on a major national, in-home, face-to-face survey of 9–19 year olds and their parents across the U.K., Livingstone and Bober (2005) show that U.K. parents are concerned that the Net may lead children to become isolated from others, expose children to sexual and/or violent images, displace more worthwhile activities and risk their privacy—concerns also widely shared by parents in China.

Nevertheless, they do value highly the potential of the Net for learning purposes. Most believe that the Internet can help their child do better at school and help them learn worthwhile things. The persistence in China of an educational system which still much reflects its historical prototype is one important indicator of the cultural continuity inherent in the theory of 'socialism with Chinese characteristics'. It thus plays a major role in framing Chinese young people's everyday lifeworlds and their relevance systems. As pragmatic individuals, urban youth are conscious of what is the 'proper' way to relate to the Internet. However, as demonstrated by the experiences of some of my informants, it may happen that the more one is expected to suppress one's expressive individualism for the sake of utilitarian goals through the exam-oriented education system, the more one tends to feel a need to escape into the virtual world for expressive needs. And the more one 'plays' online, the more one's utilitarian goals are jeopardized, so goes the (vicious) circle.

When the Net has become a mere toy or a virtual world into which one may escape from the 'real' life, its potential uses for learning (and other instrumental purposes) are naturally marginalized. Playfulness may be typical of children and young people, which can extend to their use of the Internet. However, predominantly using the Internet for fun, thus reducing it to a mere toy, may prevent one from employing it for a wider range of meaningful purposes. This is not to say that the current online recreation which Chinese youth engage in may not be meaningful and educational in some way. Like in other societies, Chinese young people's online activities may contain much informal learning as they are picking up 'new literacies' essential for participating in our contemporary and future society. Nevertheless, it seems unlikely that these skills are recognized in the Chinese context as long as the institutional structures framing young people's lives remain unchanged.

NEGOTIATING THE IDEAL CHINESE PERSONHOOD IN THE INTERNET AGE: BETWEEN EXEMPLARY NORMS AND 'THE UNEXEMPLARY'

The dramatic transformation of Chinese society has led to new, and at times confusing, contexts of identity construction for individuals and groups. This has implications, not least challenges, for governance. It is in this context that the Chinese social engineers worked out the theory of 'socialism with Chinese characteristics' and its related notion of 'two civilizations'. 'Socialism with Chinese characteristics' with its dual approach to modernity, as I have argued, presupposes a double subjectifiction regime with two sets of largely contradictory enjoinments for the subject. It incorporates both a 'do-it-yourself', 'free choice' and 'self enterprising' ethos in line with the neoliberal market economy and the enjoinment for selflessness, collectivism and plain life in line with the socialist mentality and Chinese traditional culture that tends to subordinate the individual's interest to that of the collective. The latter set of values, the Party leaders argue, will work to rectify the 'unlimited' freedom and individualism that would harm social order deemed to be the precondition for China's success in its modernization project, including constructing a 'harmonious society'. It is thus meant as a counter strategy for the former set of values which nourishes and encourages self interest and individual freedom. Therefore, there has been much effort in reinforcing this set of enjoinments through the ongoing moral education often combined with political-ideological propaganda. State-led nationalism in the form of patriotism is an important part of this process. Deliberate reinforcement is deemed necessary not least because the socialist-collectivist norms no longer make much sense for Chinese people who find themselves left in the uncharted waters of the free market economy based on the 'jungle law' without protection by the state.

The double subjectification regime combines with the notion of *suzhi* ('human quality'), whereby it is argued that the Chinese population's quality in both the earlier-mentioned aspects needs to be improved. Today the term *suzhi* is widely used in China by laypeople as well as the political and academic elite, in both formal discussions and daily life, to evaluate an individual's or a group's various attributes, ranking the latter along a scale of low and high quality. Because such a ranking has implications for one's worth, *suzhi* has been increasingly connected to the maximalization of the capabilities and value of the human body (Anagnost 2004). And as a measurement of one's worth it serves as a mechanism in the Chinese stratification system so that, in present-day China, *suzhi*, qualifications, social mobility and the 'good life' have become interrelated concepts (Liu, 2008a). In other words, *suzhi* based on the exemplary norms prescribed by the double subjectification regime has become a mechanism for rewarding and punishing, and accepting and rejecting, hence crucial for participating in the mainstream culture in the context of 'socialism with Chinese characteristics.'

The Internet has become an important site where the Chinese discourse of the ideal modern personhood is articulated, mobilized and negotiated. This is centrally reflected in the widely shared Internet anxiety. As already indicated, the Internet anxiety in China is reminiscent of the Internet addiction debates in some other countries. But it meanwhile goes far beyond the global pattern of adult concern about the Internet's allure for children and young people. It has been coproduced with larger social and political concerns concomitant to China's modernization. As an ultimate symbol of modernity, the Internet has become a site to which China's ambivalence about modernity as both signifying hope for development and danger for moral decay and the political status quo is transplanted. Thus, just as the problematization of human quality in the reform era has been accompanied by strong enjoinments for the ideal personhood of the double subjectification regime, the widespread Internet anxiety—and the accompanying anti-Internet crusade, has been accompanied by a strong official norm about the 'proper' 'netizenship' which reflects the same ideals. As a result, the norm about the 'proper' wired self reflects the qualities desired of the Chinese subject so that the exemplary netizen is at the same time the exemplary citizen, or subject, and vice versa. Young people who indulge themselves in the online world, neglecting their 'proper businesses', are often associated with being of 'low quality' in light of both the two sets of expectations of the double subjectification regime. In this way, they have failed to live up to the requirements for the loyal socialist subject as well as for filial children to their parents, with respect to both productivity and moral integrity.

The double subjectification regime means that in present-day China the new found 'freedom' is valid only to a certain degree. Chinese individuals (and groups) are supposed to be free as the economic/market and consumer subject, but not as the political and moral subject. The double regime produces a cultural condition in which even as one is embracing the 'freedom' allowed for by the former set of enjoinments to choose how to live one's life, the other set of norms is present. As such, both serve as significant cultural resources for individuals and groups to draw on in their identity work. This is reflected not only in young people's Internet-related subjectivities but also their online self-representation, whereby they display knowledge about the desired subjecthood as well as negotiate it to fit their own purposes, as I shall show in the rest of this section.

In telling about their Internet-related experiences, most of my informants, whether the 'instrumental users' or 'recreational users', volunteered a moral critique of young people and the Internet in China. This is likely a uniquely Chinese phenomenon, as it perhaps would be hard to find another society where young people are so interested in making moral judgement about the Internet and so conscious of the 'danger' of the Internet *even as they embrace it*. To negotiate a 'proper' wired self in the light of social and parental expectations seems to be imperative in the context of widespread Internet anxiety, whereby concern about the misuse of the Net is a daily discourse. The impact of the Internet anxiety on the young people is shown in the dual interpretative repertoires that they drew on in making moral evaluation of their own and other people's relationships with the Internet. According to Potter and Wetherell (1987), the descriptions and evaluations embodied in interpretative repertoires are bound together with larger cultural models of meaning making. By drawing upon existing interpretative repertoires, people talk about themselves in terms of the social order in which their lives are embedded, thus making their behavior accountable in a moral world (Shotter, 1984).

In the case of the university students' negotiation of a political self in the Internet age, although these young people have learned, by choice or circumstance, to be 'passive' in other forms of political participation, they viewed *aiguo zhuyi* (patriotism), as an important requirement for Chinese citizenship, which explains their active participation in online nationalistic activities. It is true that these young people's nationalism represents popular nationalism rather than state nationalism. Nevertheless, it still bears significant similarities and overlaps with the latter that can only be traced back to the Party state's large-scale campaign on patriotism education in the 1990s, which constituted part of the social engineering in the reform era. The post-80ers online collective narrative serves as another example of how Chinese youth display knowledge about the desired personhood. As shown in Chapter 7, in their attempts to subvert the negative stereotype of their own generation in the space allowed for by the online forum, the post-80ers tried to portray a group image of being capable of great accomplishment, which presupposes such qualities as being ambitious, adventurous, strong-willed, hardworking, entrepreneurial and responsible for one's own fate. However, they were at the same time conscious that it is not enough to be achieving individuals in the Chinese context. To further counter the negative image imposed upon their generation, they saw it necessary to demonstrate that they are not self-centered individualists concerned only about themselves, as they are often accused of. Their understanding of the 'not collapsed' thus seems to be in line with the double subjectification regime. Admittedly, this kind of 'consistency' may have to do with the purpose of the particular website, which is to counter the negative image of the post-80s generation in the Chinese media, as claimed by the website's initiators. Given the relevance of the desirable human qualities outlined by the regime for social recognition in the Chinese context, it seems natural that young people's claim for 'virtues' cannot but refer to these. With another website where much identity work is going on, for example, the blogs by Mu Zimei and Sister Furong, as mentioned earlier in this volume, one will certainly find content that is more subversive of the cultural and political conventions. Moreover, given the focus on personal life, lifestyle and consumerism, one may expect little reflection along such lines in the vast numbers of blogs in Chinese cyberspace. However, the existence of online forums such as the www.80end.com alone indicates what is at stake for young people in the age of 'socialism with Chinese characteristics.'

The analysis here points to the considerable influence on individuals' (and groups') subjectivities by the dominating discourse about what counts as desirable and acceptable human qualities in present-day Chinese society. This lends support to Stuart Hall's observation that 'identity is formed at the point where the unspeakable stories of subjectivity meet the narratives of history, of a culture' (Hall, 1987: 44). In other words, whatever we might say (and think) about ourselves and others will always be in terms of discourses provided for us by the social, cultural and historical contexts in which our lives are embedded.

However, this is not to say that the double subjectification regime has totally achieved its desired effect with young people being totally brainwashed. Rather, it is just to recognize that it constitutes one of the significant influences that one has to come to terms with. Indeed, having the knowledge and displaying it does not necessarily mean one's life is totally lived along the norms. One's lived experiences of China's (post)modernity may prove to be at odds with the exemplary norms of the double regime. Or as Bakken (2000: 98) argues, 'the "exemplary key" might not fit the "lock of market socialism."' Thus, the urban youth I interviewed were aware of the good and bad, right and wrong, useful and useless ways of interacting with the new medium, as defined in the Chinese context. Nevertheless, many of them still found it hard to live up to the ideal Chinese 'netizenship'. Moreover, they grounded their particular ways of relating to the new medium, for instance, purely for entertainment, in their own social-biographical situations characterized with pressure, frustration and boredom. Furthermore, they framed their developing a 'proper' or 'mature' relationship with the Net as a matter of coming of age, allowing themselves room and tolerance. Likewise, although the ideal personhood the post-80ers were constructing seems much in line with the desirable 'human qualities' enjoined by the double subjectification regime, their narrative is meanwhile characterized by a struggle to present a self that goes far beyond that. Alongside their claim of ideal personhood according to the social norms, the post-80ers strove to represent themselves as multi-faceted human beings, both as individuals and a group, as an effect of the dramatic social transformation that has accompanied their journey to adulthood. Furthermore, even though they talked along roughly the same double line, their notion of the ideal personhood may not be identical to the Party leaders' design. Particularly, their notion of selflessness and self-sacrifice for others may be still different from the Party state's notion of 'socialist-collectivist' values. At a closer look, one may find that in their claim of moral qualities, there is an absence of love for socialism and the Party. This is also true of my university informants' love for their Chinese national community, whereby it was emphasized that the nation is higher than the Party. Such a notion of collectivism seems more in line with the universal moral values, such as humanity and compassion for the suffering and social justice (I will return to this point in the following section.). This seems to indicate that the notion of 'revolutionary/communist' selflessness, as represented by the model figures used in China's moral education, who sacrificed themselves for the communist-socialist cause, is being replaced by a 'humanitarian selflessness', which is somewhat in line with Yan's (2009) notion of the emerging 'new sociality' in Chinese society.

Chinese youth's collective identity construction and negotiation of a 'proper' netizenship seem both to be characterized with a kind of coherence. In the case of the post-80ers, the seemingly paradoxical characteristics of Chinese urban youth, as discussed in Chapter 3, were mobilized as complementary parts of a coherent whole in the ideal form of self that they strove to represent. Likewise, the university students saw no contradiction at all between their 'political indifference' and their vehement nationalism. Rather, both were viewed as essential for the emerging modern Chinese identity that they were negotiating. When it comes to the other youth I interviewed, they drew on an interpretative repertoire of time that allowed them to see a 'proper' relationship with the Internet as an emerging process. The conflicts between the different and competing discourses are eased in all these cases—and yet, only discursively. This fits the view that identity construction is an emergent and ongoing narrative project which is meanwhile characterized by a struggle for coherence (Goodson and Sikes, 2001; Josselson, 1996). Just as Goodson and Sikes (2001: 41) have argued, although 'self' is a contested term and at a very basic level we are multi-self beings, we tend to work hard to present ourselves as having a unified coherent identity, or at least narrate our lives according to an aspiration for coherence, for a unitary self, because to fail to do this is to come across as mentally unstable. It seems that in the Chinese case, a sense of coherence relies much on a kind of 'dualism' that treats seemingly contradictory elements as complementary parts of a balanced whole, a point I will come back to at the end of this book.

'RE-COLLECTIVIZATION' IN THE CONTEXT OF INDIVIDUALIZATION: IMAGINING COMMUNITIES ONLINE

Online nationalism and online community building are two of the most striking motifs along with the rampant consumerism in Chinese cyberspace. The case studies in this volume are but two instances of these. What, then, does this popularity of collectivism mean? In other words, how can we make sense of the eagerness for collective identity construction, which is shared by the online nationalism and online community building, in a context of social fragmentation and strong individualistic tendencies? Nationalism and other forms of community building may not be contradictory to individualization. After all, even in societies known as the most individualistic ones (e.g., the U.S.), one finds both. However, when individualism prevails, community tends to be weakened, according to some scholars writing about the individualization process in the U.S. (Bellah et al., 2008; Putnam, 2000; Riesman, 1989). To a great extent, this seems also true of Chinese society in the reform era, where there has been a process of fragmentation of society as the country undergoes far-reaching individualization. On the other hand, however, it is perhaps precisely because of the fragmentation that people may feel a greater need for a sense of community and collectivism. This seems especially true in societies where the individualization process has not been accompanied by institutional support for individuals, such as in the Chinese case. The vehement (online) nationalism and the flourishing online communities in China seem to indicate that a process of 're-collectivization' has been concomitant to the process of individualization. This reflects, and reinforces as well as is reinforced by, the complexities, contradictions, multi-temporalities and hybridities of social and cultural conditions inherent in 'socialism with Chinese characteristics'; in particular, it has to do with the individualization with its Chinese characteristics.

Since the reform and opening up started, a range of interrelated factors have led to a far-reaching process of individualization in the Chinese society, such as the market economy, increased mobility, the withdrawal of the state from most of the aspects of people's private lives, widespread consumerism and globalization (Yan, 2009). As a result, individuals enjoy more mobility, choice and freedom. But, at the same time, they are enjoined to exert agency and take responsibility for their own lives. As mentioned in Chapter 1, inherent in the free market is a new subjectification regime that requires the individual to be the freely choosing, self-enterprising and self-responsible subject. This form of the self is consistent with the autonomous subject described in Giddens' (1990, 1991) theory of reflexive modernization and promoted in the subjectification regime of the liberal democratic societies (Rose, 1998). However, unlike in those societies, where the individualization process has been backed up by the welfare state, in the Chinese case, the 'untying' of the individual has been accompanied by little institutional protection and support from the state (Yan, 2009). With the state's withdrawal from the previous socialist welfare system, it also in many ways shed its responsibilities for the provision of public goods, as exemplified by the housing reform, medical reform and education reform since the mid-1990s. Thus, it is safe to say that except for the moral and political realms, where the Party state still attempts to guide people's identity formation in line with the 'spiritual civilization', Chinese people are largely left to themselves to find means of achieving their ideal lifestyles as well as material security and welfare. As a result, at the same time as they celebrate the new 'freedom', Chinese individuals share a sense of great pressure (and insecurity), as reflected in the cases covered in this volume and my study on the only-child generation (Liu, 2006a). The pressure is reinforced by the changed notion of what counts as the 'good life' based on rampant consumerism. It also has to do with high parental expectations of the only-child, which is in turn much related to the sensed insecurity in an increasingly unequal and competitive society where social welfare for the old-aged is under-developed. Moreover, the pressure is only felt more keenly when one is simultaneously addressed by both expressive individualism and economic/utilitarian individualism with equally strong force, which is another important feature of the Chinese process of individualization.

The growth of market exchange and the dismantling of collectivism and uniformity have atomized Chinese society, removing the economic basis for comradeship characteristic of the Communist China in Mao's time. The fragmentation has been reinforced by the restratification of society due to a lack of institutional support by a welfare state. Individualization without the protection of a welfare state 'has tended to polarize society in terms of economic standing and social status, instead of merely diversifying society in terms of identity construction and life politics' (Yan, 2009: 288). Furthermore, with the relentless pressure to accumulate material wealth through individual competition—often based on unequal terms due to widespread corruption, Chinese society in the reform era is thoroughly imbued with instrumentalism and commoditization (Gold, 1985; Lin, 1999). With this, social trust (and universal values) has in general declined in China in the reform era (Yan, 2009). The sociologist Peng (2003: 292–295) claims that there are various kinds of distrust currently prevailing in Chinese society: distrust of the market due to faulty goods and bad services, of service providers and strangers, of friends and even relatives, of law enforcement officers, of the law and legal institutions and of basic moral values. This constitutes a significant feature of the Chinese 'risk society'.

The individualization and fragmentation of Chinese society has implications for political participation. The widely observed 'apoliticalness' among Chinese youth has to do with the political authoritarianism and with the fact that as urban middle class, they are basically supportive of the status quo. However, some commentators (Damm, 2007; Rosen, 2009; Yan, 2009) have also argued that 'apoliticalness' could also be a result of the individualization and fragmentation of Chinese society which has led to the shifting focus of self-development among young people from public life in the Maoist era to private life, individual lifestyles and consumerism in the reform era. Moreover, political participation, such as voting and mobilizing people to confront the official becomes more treacherous when people do not trust each other and when there is a widespread distrust of institutions due to the corruption of the power holders and the individual manipulation of the law and regulations through *guanxi* practice (Yan, 2009). This latter explanation concurs with my university informants' understanding of political participation in the Chinese context, such as their distinction between 'participation' and 'effective participation' and the perceived political inefficacy, which encourages the public mentality of 'keeping to oneself' or 'mind your own business' now that you cannot make a change.

The fragmentation of society and lack of trust in China tends to reinforce a widely shared belief in individual efforts and the understanding that one's fate lies in one's own hands (Liu, 2006a). This kind of understanding of the self has important implications for the individual and society. On the one hand, it may urge people to work hard, which has been much promoted in the Chinese culture both in the past and today with its strong belief in hard work and determination (Liu, 2002b). Few people would deny that industry is necessary for development. However, on the other hand, it reinforces a highly individualized approach to the changing reality, whereby responsibility for success or failure is placed solely on the individual in a society where people's life chances remain highly structured at the same time as they increasingly are urged to seek solution on an individual, rather than a collective basis. From a Marxist point of view, this would constitute a 'false consciousness' similar to what Furlong and Cartmel (1997: 2) refer to as an 'epistemological fallacy' inherent in the 'reflexive modernity' thesis:

> Although social structures, such as class, continue to shape life chances, these structures tend to become increasingly obscure as collectivist traditions weaken and individualist values intensify. As a consequence of these changes, people come to regard the social world as unpredictable and filled with risks which can only be negotiated on an individual level, even though chains of human interdependence remain intact.

According to Furlong and Cartmel, risks, uncertainty and pressure are the consequences of the very pressures to adopt individualistic perspectives in a society characterized by interdependency.

It is here I see a connection between Chinese young people's online collective narratives, be it in the form of community building or popular nationalism, and the individualization process in China. Online nationalism and community building can both be seen as forms of, or attempts at, a 're-collectivization' in the context of the particular individualization process of Chinese society characterized with great pressure, insecurity and lack of social trust. Online nationalism and online community building in this sense represent an awareness of the aforementioned 'epistemological fallacy' among Chinese young individuals, who may have come to realize, based on their lived experiences of China's modernization, that after all they need a community to fall back on. Within such a context, collective identity construction and nationalism, both online and offline, allow for a sense of belonging, solidarity, trust and community—hence dignity, which is often lacking in present-day Chinese society. Or in Yang's (2009b: 156) words, they both represent a 'utopian impulse' which is 'a critique of the present and a yearning for a better world. It originates in the social, cultural, and political displacements associated with the market transformation.'

To be sure, the Chinese Party state has also attempted to 're-collectivize' Chinese society in the face of the individualization and fragmentation by means of a double subjectification regime that enjoins its subjects to embrace the socialist-collectivist ideals as well as the neoliberal norms. This top-down 're-collectivization' strategy, however, differs from the bottom-up attempts at 're-collectivizing' by Chinese individuals, such as the cases covered in this volume. It is still based on a negation of the self and the subordination of individual interest to that of the collectives defined by the Party state, which include the people, the socialist state and the Party. Such a value system has so far largely failed to appeal to today's young people. In contrast, although online communities and nationalism also emphasize the interrelatedness between the 'collective' and the 'individual', there is at the same time a sense of equality and democratic participation among the members of the 'imagined' communities, be it the online forum or the nation. Both allow common people to join with others in forming ideals of equality and solidarity, hence dignity in a context characterized by inequality, lack of trust and patriarchy. As such, these forms of collective identity construction may be especially valuable in a society where both 'cultural democracy', which denotes the wide acceptance and practice of democracy as a principle in everyday life and social relations, and political democracy are still far from being a reality (Yan, 2009: 288). As such, online community and nationalism, as reflected by the cases discussed in this volume may serve as a fulfillment of the otherwise unavailable democratic public life in China. In socio-cultural contexts such as China, where space for such collective narratives is rather limited offline, the Internet proves to be a valuable extension or compensation for the offline reality. The new technology makes it possible (at least potentially) for geographically and socially isolated individuals to engage in collective narratives on an even larger scale and density compared with the traditional media.

A POSTSCRIPT: CHINESE YOUTH AND DUAL MODERNITY

In this book I use the concept of 'dual modernity' to denote the Chinese state's approach to modernization—exemplified in the twin emphasis on economic development and social control and the concomitant double subjectification regime. Dualism (or just duality) is also useful for understanding Chinese young netizens' identity construction. To be sure, although youth's dualistic approach to identity reflects and overlaps to a certain extent with that of the state, the two are not identical. Nevertheless, it would seem that coping with apparent dual*ism* (between contradictory forces and elements) and making it into a 'duality' (which does not necessarily denote contradiction), by even making it *a virtue* is a Chinese characteristic both at the individual level and at the level of state policy. Apparent contradictions are of course not a uniquely Chinese characteristic, neither at the level of individual (and group) identity construction, nor at the level of government policy. But, it seems to be very Chinese to make a virtue of duality so that seemingly contradictory elements appear not as conflicting but as complementary components of a harmonious balance. Could it be, then, that what may appear contradictory or paradoxical from a Western perspective, need not appear to be so at all from a Chinese point of view?

At the end of the 1970s, China opened itself up to the world, hoping to escape from the economic collapse and political chaos caused by Mao's social movements. It has ever since found itself in the current of two forces—modernization at home and globalization in the world. Seeking, on the one hand, to realize the hope of rapid economic development under conditions of globalization, and on the other hand, to avoid the dangers which capitalist globalization posted for political and cultural control, the Party leaders invented China's own modernization theory, epitomized in the notion of 'socialism with Chinese characteristics'. This theory of 'socialist modernization' involves a hybridization of two seemingly contradictory ideologies: economic liberalism and political authoritarianism. Such a dualism underlies the notion of 'two civilizations' as a further development of 'socialism with Chinese characteristics'. By this, the Party leaders insist that socialist modernization depends on both 'socialist spiritual civilization construction' and 'socialist material civilization construction', respectively representing economic development and social control. China's 'dual modernity' has entailed a double subjectification regime, which requires the newly liberated Chinese subject to cultivate human qualities along the lines of the 'two civilizations'. Such a dualism represents the state's search for a modern identity for China.

Just as the current Chinese state is seeking for a unique modern identity, China's younger generation is also displaying a much stronger desire than earlier generations for a distinct cultural identity that marks their local, regional, ethnic and national differences. Young people's identity work, however, has to be done within a context of unprecedented complexity and apparent confusion due to the various 'glocal' forces concomitant to 'socialism with Chinese characteristics'. Under such circumstances, Chinese youth's identity construction revolves around a series of apparent paradoxes. They come across as simultaneously expressive and instrumental, radical and pragmatic, modern and traditional, individualistic and collectivist, hedonistic and 'puritanical', materialistic and idealistic, romantic and realistic, internationalist and nationalist, global and local, apolitical and yet nationalistic, blessed with material prosperity and yet under great pressure to strive for greater material wealth and exposed to an unprecedented range of choices and yet lacking a sense of security.

Young people's dualistic approach to identity construction may well have been influenced by government policy to some extent. However, it would be overly simplistic to assume that members of the younger generation are the carbon copies of what the Party leadership intends to produce by means of the double subjectification regime. Rather, their dualism takes the form of 'multiple dualities' that cannot be reduced to the state's dualism. On the one hand, youth's dualistic approach represents their attempts at coming to terms with the moral order of their society drawing on the larger cultural models of meaning making. On the other hand, the wide range of dual elements on which they draw in fashioning their individual and group identities represent their response to the multiple and competing discourses available in this age of 'glocalization'. The multiple cultural influences addressing today's Chinese youth go far beyond the dual qualities enjoined by the double subjectification regime. Youth's multiple dualism, thus, may challenge the political and cultural conventions at the same time as it may perpetuate them.

Clearly, Chinese youth's approach to identity construction is highly eclectic. Hybrid or eclectic approaches to the self, however, are by no means uniquely Chinese, but apply also to youth in other societies. In this age of 'global modernity', there has been a new emphasis on reflexivity—self-conscious choice in the shaping of identities (Nilan and Feixa, 2006). Nilan and Feixa (2006) claim that all youth today engage in such reflexivity to a greater or lesser degree. In so doing, youth negotiate between apparently competing identity discourses on offer at all the local, regional and global levels by filtering, mixing, synthesizing and judicious choice. It is not uncommon that youth in the South combine Western consumerism with traditional values, be it religious beliefs or family cultures. In short, it seems that socio-cultural hybridity, creative synthesis and pluralistic identities have become common to young people in fashioning a modern self. Moreover, what would be contradictory identity discourses to an older generation often do not seem so to youth, who pull upon a pastiche of sources in their creative practices (Willis, 1990).

What gives the story about Chinese youth a cultural turn, however, is that their 'reflexive' construction of the self is characterized by a pairing of a series of seemingly contradictory elements. In each pair, the two elements are employed as mutually *complementary* and *interdependent* rather than as conflicting. This rests easily with Chinese traditional dialectics, the major element of which is the notion of the 'permeation of opposites' or 'complementary opposition' (*xiangfan xiangcheng*) (Tian, 2005). Such patterns of thinking sought 'to merge seemingly conflicting elements into a unified harmony' (Meisner, 1977: 257). It is widely known that there is a strong emphasis on harmony in the Chinese tradition. And according to the Chinese style of reasoning, harmony is achieved only by striking a balance between seemingly conflicting elements, such as *Yin* and *Yang*, heaven and earth, male and female, light and dark, strong and weak, father and mother. Seemingly contradictory elements are strategically employed in complementary pairs in the hope of achieving a certain effect—balance, hence harmony. The emphasis on balance is also reflected in the notion of '*zhongyong*' (the Doctrine of the Mean), which advocates avoidance of excesses or extremes. The goal of the mean is to maintain balance and harmony by directing the mind to a state of constant equilibrium. Between contradictory elements, one is expected to stand at the middle point. This represents the Chinese approach to solving conflicts and contradictions: not by eliminating them, but by coupling the two apparently contradictory elements into a unified one in a balanced way.

Expressions such as 'paradoxes', 'contradictions' or 'conflicts' have been frequently used in this book (and indeed in other academic discussions as well) to make sense of Chinese ways of thinking and doing both at the state level and the individual and group levels in the era of 'dual modernity' under 'socialism with Chinese characteristics'. Seen through a Chinese lens, however, such expressions may be valid only from the perspective of outsiders, including researchers who see the world through Western cultural lenses. For Chinese youth, like for the Party leaders, both of whom show a striking dualism in negotiating modern identities, what others view as 'paradoxes' or 'contradictions' may be just pairings of two complementary elements, both of which are necessary for a sense of balance and harmony.

However, it is also possible that this discrepancy may be because we as researchers access their realities mainly through *their* story-telling, at least in this case. Like other people's self narratives, the Chinese ones are also necessarily characterized with an effort to strike coherence. As mentioned earlier on, struggling for a sense of unity and coherence, which is typical of identity construction, stands for attempts at coping discursively with the complex and at times confusing human conditions. It is just that, for the Chinese, as reflected by the youth in this book, it seems that the coherence of the self relies much on a dualistic approach, namely, the coupling of seemingly contradictory elements into a unified one by using them complementarily. By the same token, the Chinese dualistic approach to modernity, be it at the state level or the individual or group levels may not nullify the difficulties, contradictions, dilemmas and confusions of the social realities that pose crisis in 'identity politics'. In this book, a strong sense of pressure, frustration and boredom constitute a persistent theme in youth's Internet-related perceptions and experiences and their online self-representation.

In summary, it seems that for Chinese youth, as for the Party leaders, dualism represents a pragmatic response to the conflicting cultural demands posed by today's 'glocal' conditions. It also stands for the Chinese ideal for human beings and society, which is as much informed by the Chinese traditional culture of harmony as by the intensified complexities of human conditions in our contemporary world. It, thus, represents a 'cultural innovation' within a tradition. For Chinese today, more than ever before, negotiating identity becomes an art of fine balance between all the competing and often contradictory cultural influences, and between dreams and realities. And as in an art of fine balancing, the balance is not always easy to keep.

Notes

NOTES TO THE INTRODUCTION

1. A range of expressions have been used for Chinese society since the reform, such as 'post-reform China', 'post-socialist China', 'late-socialist China', 'post-Mao China', 'reform-era China', 'the reform era' and 'the reform and opening up era'. Such expressions are used interchangeably in this book.
2. China Internet Network Information Center (CNNIC) is China's officially assigned Internet research and administrative agency. It conducts a nation-wide survey twice a year on the development of the Chinese Internet. Two points are worth mentioning about CNNIC. First, its semi-annual surveys are useful in revealing the general development patterns of the Chinese Internet, but they do not provide more nuanced data on uses and users. Secondly, although official statistics should be read with a critical eye, the rapid growth of China's Internet is indisputable. CNNIC's reports are all easily available at http://www.cnnic.org.cn, in both English and Chinese.
3. The separation of rural and urban sectors has been a major feature of Chinese society. Despite Mao's slogan of 'abolishing the urban-rural disparities', his urban-centered policies after the founding of the People's Republic of China increased these disparities. Among the policy measures, restrictions on rural-to-urban migration, largely by means of the residency permits system, known as *hukou*, made it possible to maintain low urban consumption and to increase industrial investments. Prior to the start of the reform in 1978, capital goods were excessively concentrated in the urban sector. Urban dwellers enjoyed a range of social, economic and cultural benefits whereas China's rural residents, the majority of the population, were treated as second-class citizens. Documentation of the origin and features of the Chinese dualistic economic structure can be found in Rawski (1979), Lardy (1983), Perkins and Yusuf (1984), Nolan and White (1984) and Riskin (1987).
4. The workshop was held on October 2–3, 2003. See http://nias.ku.dk/.
5. See F. S. Liu (2004).

NOTES TO CHAPTER 1

1. I use the Chinese yuan or U.S. dollars in this book depending on usage in the source cited.
2. *Xia Hai,* literally, 'jumping into the sea', is an expression widely used in the reform era to denote the fever of giving up state-secured jobs for self-employment

and private businesses (especially in the early 1990s) in the hope of greater material benefits and individual freedom. The term has become a metaphor for both the broadened scope of choice and increased risk for the individual in the reform era.

NOTES TO CHAPTER 2

1. Information about the earlier development of China's Internet is obtained mainly from CNNIC's homepage. See: *The Internet Timeline of China* at http://www.cnnic.cn/en/index/0O/06/index.htm. I also consulted other authors on this issue, e.g. Zhou (2006) and Qiu (2003).
2. See note 1.
3. Information about the number of China's cyber police seems to vary from source to source. For example, Lum (2006) estimates it to be 3,000 whereas Mooney (2004) claims it to be 4,000.
4. See http://ciirc.china.cn/.
5. This phenomenon is recorded at http://en.wikipedia.org/wiki/Harmonious_society. One also encounters such usages by Chinese Internet users on a regular basis.
6. More of Jiang's writing about the importance of informatization can be found in his prologue for the book *Exploration and Practice for Informatization in China*, Qili Hu (ed.) (2001) Beijing: Publishing House of Electronics Industry.
7. Global Technology Forum (November 9, 2005) *Will China's youth play virtuous virtual game?* (http://globaltechforum.eiu.com/index.asp?layout=rich_story&doc_id=7780&categoryid=&channelid=&search=communicating).
8. See http://homes.eff.org/~barlow/Declaration-Final.html.
9. ResearchinChina (2008) China Online Games Market Report 2007–2008 (http://www.companiesandmarkets.com/Summary-Market-Report/china-online-games-market-report-2007-to-2008–47986.asp).
10. See http://chinadigitaltimes.net/2009/06/xiao-qiang-the-roar-of-dissent-online/.
11. See http://chinadigitaltimes.net/2010/01/southern-metropolis-weekly-top-10-neologisms-of-2009-part-i/.

NOTES TO CHAPTER 3

1. The 2005 nationwide survey and the statistics from the China Psychology Association are cited by Yan (2006). I have failed to find the original sources of these, as they are neither available in Yan's article, nor on the Internet. This applies also to the 2004 survey of 1,010 college students in the city of Wuhan and the survey of 200 students in a Beijing college to be mentioned in the following pages.
2. See http://www.for68.com/new/2009%5C11%5Cxi910732404311111900214364–0.htm or http://www.shanghaidaily.com/buzzword/article.asp?id=202.
3. 'Ku' is written with a classical Chinese character 酷 (also pronounced ku) whose original meaning was 'cruel'. This written character is now more commonly employed to represent the new slang term in current popular culture contexts rather than the classical word ku (Moore, 2005).
4. See http://baike.baidu.com/view/8234.htm

5. This is from my project on the first-generation only-children's identity construction (see Liu, 2006a). However, this part of the interview was not reported in the published work.
6. See http://zhidao.baidu.com/question/7137518.html.

NOTES TO CHAPTER 5

1. This quote is taken from an article included in zhidao.baidu.com ('Baidu knows') with the title 'What harms does youth's addiction to the Internet cause?' which is based on Tian Bingxin's interview with Zhang. See: http://zhidao.baidu.com/question/8753706.html?si=1. Also see Tian Bingxin (2005). *Zhang Chunliang: Damned Internet Games.* At http://www.southcn.com/nfsq/ywhc/tbxst/shentan/200510260380.htm.
2. See http://chinaonmeta.com/news?nid=11247.
3. *A Global Recovery* (March, 23, 2009). China Implements New Anti-Internet Addiction System (http://aglobalrecovery.com/2009/03/23/china-implements-new-anti-internet-addiction-system/).
4. Ariana Eunjung Cha, Washington Post (February 21, 2007) Strict clinic near Beijing is operated by a military researcher (http://www.washingtonpost.com/wpdyn/content/article/2007/02/21/AR2007022102094.html).
5. See http://game.people.com.cn/GB/48644/48662/4122463.html.
6. Reports about China's measures against Net addiction, both hard and soft, can be found at: http://chinadigitaltimes.net/china/internet-addiction/.

NOTES TO CHAPTER 6

1. This is taken (and translated) from what seems to be a widely circulating conversation between Tian Bingxin and Zhang Chunliang. See note 1 in Chapter 5.

References

Anagnost, A. (2004) The corporeal politics of quality (*suzhi*). *Public Culture* 16(2): 189–208.

Anderson, B. (1991) *Imagined communities: Reflections on the origin and spread of nationalism*. London: Verso.

Appadurai, A. (1996) *Modernity at large: Cultural dimensions of globalization*. Minneapolis, MN: University of Minnesota Press.

Armstrong, S. and Warlick, D. (2004) The new literacy: The 3 Rs evolve into the 4 Es. *Technology & Learning* 25(2): 20–24.

Arnett, J. J. (2000) Emerging adulthood. *American Psychologist* 55: 469–480.

Aune, M. (1996) The computer in everyday life: Patterns of domestication of a new technology. In *Making technology our own? Domesticating technology into everyday life*, M. Lie and K. H. Sorensen, eds., 91–120. Oslo: Scandinavian University Press.

Bakardjieva, M. (2005) *Internet society: The Internet in everyday Life*. London: Sage.

Baker, P. (2001) Moral panic and alternative identity construction in Usenet. *Journal of Computer Mediated Communication* 7(1): 1–1. At: www.ascusc.org/jcmc/vol7/issue1/baker.html

Bakken, B. (2000) *The exemplary society: Human improvement, social control, and the dangers of modernity in China*. Oxford, UK: Oxford University Press.

Bamberg, M. (2005) Narrative discourse and identities. In *Narratology beyond Literary Criticism*, J. C. Meister, T. Kindt, W. Schernus, and M. Stein, eds., 213–237. Berlin and New York: Walter de Gruyter.

Bandurski, D. (2008) China's Guerilla War for the Web, *The Far Eastern Economic Review*, July 2008. At: http://www.feer.com/essays/2008/august/chinas-guerrilla-war-for-the-web?

Barnes, S. H. and Kaase, M. (1979). *Political action: Mass participation in five western democracies*. New York: Sage.

Beavis, C., Nixon, H. and Atkinson, S. (2005) LAN cafés: Cafés, places of gathering or sites of informal teaching and learning? *Education, Communication and Information* 5(1): 41–60.

Bellah, R. N., Maden, R., Sullivan, W. M., Swidler, A. and Tipton, S. M. (2008) *Habits of the heart: Individualism and commitment in American life*. Berkeley, CA: University of California Press.

Benkler, Y. (2006) *The wealth of networks: How social production transforms markets and freedom*. New Haven, CT: Yale University Press.

Bian,Y. J. (2002) Chinese social stratification and social mobility. *Annual Reviews of Sociology* 28: 91–116.

Block, J. J. (2008) Issues for DSM-V: Internet addiction. *American Journal of Psychiatry* 165(3): 136–37.

Boyd, D. (2008) Why youth (heart) social network sites: The role of networked publics in teenage social life. In *Youth, Identity, and Digital Media*, D. Buckingham, ed., 119–142. Cambridge, MA: MIT Press.
Bray, D. (2005) *Social space and governance in urban China: The Danwei system from origins to reform*. Palo Alto, CA: Stanford University Press.
Breuilly, J. (1993 /1982) *Nationalism and the state*. Chicago, IL: University of Chicago Press.
Brown, P. (2003) The opportunity trap: Education and employment in a global economy. *European Educational Research Journal* 2(1): 141–179.
Brown, A. D. and Humphreys, M. (2002) Nostalgia and the narrativization of identity: a Turkish case study. British Journal of Management 13 (2): 141–159.
Brubaker, R. and Cooper, F. (2000) Beyond 'identity.' *Theory and Society* 29(1): 1–47.
Bryson, M., MacIntosh, L., Jordan, S. and Lin, H. (2006). Virtually queer? Homing devices, mobility, and un/belongings. *Canadian Journal of Communication* 31: 791–814.
Bucher, H. J. (2004) Is there a Chinese internet? Intercultural investigation on the Internet in the People's Republic of China: Theoretical considerations and empirical results. In *Fourth international conference: Cultural attitudes towards technology and communication 2004 proceedings*, Fay Sudweek and Charles Ess, eds., 416–428. Murdoch, Australia: Murdoch University.
Buckingham, D. (2003) *Media education: Literacy, learning and contemporary culture*. Cambridge, UK: Polity.
———. (2006) Children and new media. In *Handbook of new Media: Social Shaping and social consequences of ICTs* (Updated student edition), L. A. Lievrouw and S. Livingstone, eds., 75–91. London: Sage.
———. (2008) Introducing identity. In *Youth, identity, and digital media*, David Buckingham, ed., 1–24. Cambridge, MA: MIT Press.
Budgeon, S. (2003). *Choosing a self: Young women and the individualization of identity*. Westport, CT: Praeger.
Burn, A. and Willett, R. (2004) *What exactly is a paedophile?: Children talking about Internet risk*. London: Centre for the Study of Children, Youth and Media Institute of Education, University of London. At: http://www.ccsonline.org.uk/mediacentre/Research_Projects/paedophilia_article2.htm
Calhoun, C. (1991). Indirect relationships and imagined communities: Large-scale social integration and the transformation of everyday life. In *Social theory for a changing society*, P. Bourdieu and J. Coleman, eds., 95–121. New York: Sage.
Candelaria, C., Daly, M. and Hale, G. (2009) *Beyond Kuznets: Persistent regional inequality in China*. San Francisco: Federal Reserve Bank of San Francisco, Working Paper Series. At: http://www.frbsf.org/publications/economics/papers/2009/wp09–07bk.pdf (Accessed January 2, 2010).
Cao, Y. and Downing, J. D. H. (2008) The realities of virtual play: Video games and their industry in China. *Media Culture Society*, 30(4): 515–529.
Casid, J. H. (2005) *Sowing empire: Landscape and colonization*. Minneapolis, MN: University of Minnesota Press.
Chan, B. (2005) Imagining the homeland: The Internet and diasporic discourse of nationalism. *Journal of Communication Inquiry*, 29(4): 336–368.
Chan, K., and McNeal, J. U. (2003) Parental concern about television viewing and children's advertising in China. *International Journal of Public Opinion Research* 15: 151–66.
———. (2004) *Advertising to children in China*. Hong Kong: The Chinese University Press.

Chang, H. (1987) *Chinese intellectuals in crisis: Search for order and meaning, 1890–1911*. Berkeley, CA: University of California Press.
Chase, M. and Mulvenon, J. C. (2002) *You've got dissent: Chinese dissident use of the Internet and Beijing's counter-strategies*. Santa Monica, CA: RAND.
Chee, B. W. L. (2000). Eating snacks and biting pressure: Only children in Beijing. In *Feeding China's little emperors: Food, children and social change*, J. Jun, eds., 48–70. Stanford: Stanford University Press.
Chen, S. B. (2005) Jiegou zhongguo chengshi hexin jiating haizi xiaofei. Deconstructing children's consumption in Chinese urban nuclear families. Guoji Guanggao. *International Advertising* 6 (June): 67–70.
China Daily (May 29, 2003) *Wired generation finds wires crossed*. At: http://www.chinadaily.com.cn/en/doc/2003–05/29/content_167045.htm (accessed June 20, 2006).
———. (October 8, 2003) *Digital era dawns at Olympics*. At: www.chinadaily.com.cn (accessed July 29, 2006).
———. (February 18, 2005). *Middle class society a long way off in China*. At: http://www.chinadaily.com.cn/english/doc/2005–02/18/content_417241.htm (accessed January 20, 2006).
———. (April 1, 2005) *Parents pay too much for education*. At: http://www.chinadaily.com.cn (accessed October 6, 2009).
———. (June 19, 2005) *Income gap in China widens in first quarter*. At: http://www.chinadaily.com.cn/english/doc/2005–06/19/content_452636.htm (accessed April 7, 2006).
———. (January 20, 2007) *90m Chinese grow up as 'only' children*. At: http://www.chinadaily.com.cn/china/2007–01/20/content_788315.htm (accessed January 5, 2010).
———. (July 8, 2008) *The war between the 80s and the 90s in China*. At: http://www.chinadaily.com.cn/china/2008–07/08/content_6827911.htm (accessed February 10, 2010).
———. (July 2, 2009) *Communist Party retains vigor, charisma among Chinese youth 88 years after its founding*. At: http://english.peopledaily.com.cn/90001/90776/90785/6691284.html (accessed May 5, 2010).
———. (January 22, 2010). *Employment rate of graduates hit 87% in 2009*. At: http://english.sina.com/china/2010/0121/301371.html (accessed June 5, 2009).
China Internet Network Information Center (CNNIC) (2007) *China's blogger survey report* [in Chinese]. China Internet Network Information Center. At: www.cnnic.net.cn/uploadfiles/pdf/2006/9/28/182836.pdf (accessed March 23, 2009).
———. (2008) *Survey Report on Internet Development in China*. Beijing, China. At: http://www.cnnic.org.cn (accessed June 5, 2009).
China Statistical Bureau (2004) *China statistical yearbook*. Beijing: China Statistics Press.
China Internet Network Information Center (CNNIC (2009) Survey report on Internet development in China. Beijing, China. At http://www.cnnic.org.cn (accessed August 2, 2009).
China Internet Network Information Center (CNNIC (2010) Survey report on Internet development in China. Beijing, China. At At http://www.cnnic.org.cn (accessed August 2, 2010).
———. (2010) *China statistical yearbook*. Beijing: China Statistics Press.
Conge, P. J. (1988) The concept of political participation: Toward a definition. *Comparative Politics* 20(2): 241–249.
Connelly, R. and Zheng, Z. Z. (2003) Determinants of school enrolment and completion of 10 to 18 year olds in China. *Economics of Education Review* 22(4): 379–388.

Croll, E. J. (1995) *Changing identity of Chinese women: Rhetoric, experience and self-Perception in the twentieth-century China*. Hong Kong: Hong Kong University Press.

———. (2006) The Intergenerational Contract in the Changing Asian Family. *Oxford Development Studies* 34(4): 473–491.

Cullinane, J. (2003) “Net”-working on the Web: Links between Japanese HIV patients in cyberspace. In *Japanese cybercultures*, N. Gottlieb and M. McLelland, eds., 126–140. London: Routledge.

Dai, X. D. (2003) ICTs in China’s development strategy. In *China and the Internet: Politics of the digital leap forward*, C. R. Hughes and G. Wacher, eds., 8–29. London: Routledge.

Damm, J. (2007) The Internet and the fragmentation of Chinese society. *Critical Asian Studies* 39(2): 273–294.

Davies, B. and Harre, R. (1990) Positioning: The discursive production of selves. *Journal for the Theory of Social Behaviour* 20(1): 43–63.

Davis, D. (2005) Urban consumer culture. *The China Quarterly* 183, 692–709.

Deane (1992) The little emperors. *Los Angeles Times Magazine*. July 26, 17–22.

Deng, X. P. (1983) *Constructing socialism with Chinese characteristics [Jianshe you zhongguo tese de shehui zhuyi]*. Beijing: People’s Press.

Deutsch, F. M. (2006) Filial piety, patrilineality, and China’s one-child policy. *Journal of Family Issues* 27(3), 366–389.

DiMaggio, P. and Hargittai E. (2001) From the ‘digital divide’ to ‘digital inequality:’ Studying Internet use as penetration increases. Princeton University Center for Arts and Cultural Policy Studies, Working Paper Series number 15. At: www.webcitation.org/5Nwy95r09 (accessed June 28, 2010).

Dutton, M. (1992) *Policing and punishment in China: from patriarchy to ‘the people’*. Cambridge, UK: Cambridge University Press.

Eagleton, T. (1997) The contradiction of postmodernism. *New Literay History* 28(1): 1–6.

Edmunds, J. and Turner, B. (2002). *Generations, culture and society*. Buckingham, UK: Open University Press.

Elegant, S. (July 26, 2007) China’s me generation. At: http://www.time.com/time/magazine/article/0,9171,1647228,00.html (accessed May 15, 2009).

Elisabeth J. F. (2007) Lesbians in (cyber)space: the politics of the internet in Latin American on- and off-line communities. *Media, Culture & Society* 29(5): 790–811.

Ellison, N. B., Steinfield, C. and Lampe, C. (2007) The benefits of Facebook “friends”: Social capital and college students’ use of online social network sites. *Journal of Computer-Mediated Communication* 12(4): 1143–1168.

Facer, K., Furlong, J., Furlong, R. and Sutherland R. (2001) Constructing the child computer user: From public policy to private practices. *British Journal of Sociology of Education* 22(1): 91–108.

Fan, S. G., Kanbur, R. and Zhang, X. B. (2009) *Regional inequality in China: trends, explanations and policy responses*. New York: Routledge.

Farquhar, J. and Zhang, Q. C. (2005) Biopolitical Beijing: Pleasure, sovereignty, and self-cultivation in China’s capital. *Cultural Anthropology* 20(3): 303–327.

———. (2002) Opening Up: Youth Sex Culture and Market Reform in Shanghai. Chicago: The University of Chicago Press.

Farrer, J. (2005) Interpreting the ‘Mu Zimei phenomenon’: Chinese sexual politics in the Internet age. Conference presentation, at the international conference of Asia scholars (ICAS 4), Shanghai.

Festini F. and de Martino, M. (2004) Twenty five years of the one-child family policy in China, *Journal of Epidemiology and Community Health* 58: 358–360.

Fletcher, G. (1998) . . . All that is virtual bleeds into reality. At: www.spaceless.com/papers/1.htm (accessed January 28, 2003).

Fluck, W. (1998) The Humanities in the age of expressive individualism and cultural radicalism. *Cultural Critique, No. 40, The futures of American studies*, 49–71.

Flurry, L. A. and Veeck, A. (2009) Children's relative influence in family decision making in urban China. *Journal of Macromarketing* 29(2): 145–159.

Fong, V. L. (2004a) *Only hope: Coming of age under China's one-child policy.* Palo Alto, CA: Stanford University Press.

———. (2004b) Filial nationalism among Chinese teenagers with global identities. *American Ethnologist* 31(4), 631–648.

———. (2007) Morality, cosmopolitanism, or academic attainment? Discourses on "quality" and urban Chinese-only-children's claims to ideal personhood. *City & Society* 19(1): 86–113.

Foucault, M. (1982) The subject and power. In *Michel Foucault: Beyond Structuralism and Hermeneutics*, H. Dreyfus and P. Rabinow, eds., 208–228. Brighton, Sussex, UK: Harvester Press.

———. (1984) Space, knowledge and power. In *The Foucault reader*, P. Rabinow, ed., 239–256. New York: Pantheon Books.

———. (1986) Of other spaces. *Diacritics* 16(1): 22–7.

———. (1990) *The history of sexuality, 1, An introduction.* New York: Vintage Books.

———. (1997) *Ethics: Truth and subjectivity: The essential works of Foucault 1954–1984.* New York: The Free Press.

Franklin, M. I. (2003) I define my own identity: Pacific articulations of 'race' and 'culture'on the Internet. *Ethnicities* 3(4): 465–490.

Furlong, A. and Cartmel, F. (1997) *Young people and social change: Individualization and risk in late modernity.* Buckingham, UK: Open University Press.

Gamble, J. (2003) *Shanghai in transition: Changing perspectives and social contours of a Chinese metropolis.* London: Routledge.

Gao, L. B. (1998) Cultural context of school science teaching and learning in the People's Republic of China. *Science Education* 82(1): 1–13.

Garnaut, R. (2008) *Thirty years of Chinese reform and economic growth: Challenges and how it has changed world development.* Paper presented at Thirty Years of Reform and Development in the PRC: Retrospect and Prospects CCER, Peking University, Beijing, October 25–26. At: http://rspas.anu.edu.au/economics/publish/papers/garnaut/ThirtyYearsChineseReform_251008v2.pdf (accessed November 9, 2009).

Gauntlett, D. (2010/forthcoming) *Making is Connecting.* At: http://www.makingisconnecting.org/gauntlett2010-extract2.pdf (accessed March 9, 2009).

Gee, J. P. (2000) The new literacy studies: Form "socially situated" to the work of the social. In *Situated literacies: Reading and writing in context*, D. Barton, M. Hamilton and R. Ivanic, eds., 180–96. London: Routledge.

———. (2003) *What video games have to teach us about learning and literacy.* New York: Palgrave.

Geis, J. P. and Holt, B. (2009) "Harmonious society": Rise of the new China. *Strategic Studies Quarterly* Winter: 75–94.

Georgiou, M. (2002) Diasporic communities online: A bottom up experience of transnationalism, *Hommes et Migrations* 12(40): 10–18.

Ghose, A. K. (2005) Employment in China: Recent trends and future challenges. At: http://www.ilo.org/public/english/employment/strat/download/esp2005-14.pdf (accessed February 17, 2006).

Giddens, A. (1990). *The consequences of modernity.* Cambridge, UK: Polity.

———. (1991). *Modernity and self-identity: Self and society in the late modern age.* Cambridge, UK: Polity.

———. (1994). Living in a post-traditional society. In *Reflexive modernization*, U. Beck, A. Giddens, and S. Lash, eds., 56–109. Cambridge, UK: Polity.
Giese, K. (2006). *Challenging party hegemony: Identity work in China's emerging virreal places*. Paper in GICA (German Institute of Global and Area Studies) Working paper series. No.14. At http://ideas.repec.org/p/gig/wpaper/14.html (accessed September 2008).
Giles, J., Park, A. and Cai, F. (2006) *How has enterprise restructuring affected China's urban workers? The China Quarterly* 177: 61–95.
Global Technology Forum (November 9, 2005) *Will China's youth play virtuous virtual game?* At: http://globaltechforum.eiu.com/index.asp?layout=rich_story&doc_id=7780&categoryid=&channelid=&search=communicating (accessed March 4, 2010).
Global Times (July 16, 2009) Taking the virtual plunge with online marriage games. At: http://www.globaltimes.cn/www/english/sci-edu/china/2009–07/447818.html (accessed January 10, 2010).
Gold, T. B. (1985) After comradeship: Personal relations in China since the cultural revolution. *The China Quarterly* 104: 657–75.
Golub, A. and Lingley, K. (2008) Just like the Qing empire: Internet addiction, MMOGs, and Moral crisis in contemporary China. *Games and Culture: A Journal of Interactive Media* 3(1): 59–75.
Gong, W., Li, Z. G. and Li, T. (2004) Marketing to China's youth: A cultural transformation perspective. *Business Horizons* 47:41–50.
Goodson, I. F. and Sikes, P. (2001) *Life history research in educational settings: Learning from lives*. Buckingham, UK: Open University Press.
Grant, B. M. (2005) Fighting for space in supervision: Fantasies, fairytales, fictions and fallacies'. *International Journal of Qualitative Studies in Education* 18(3): 337–354.
Greenfield, L. (1992) *Nationalism: Five roads to modernity*. Cambridge, MA: Harvard University Press.
Greenhalgh, S. and Winckler, E. A. (2005) *Governing China's population: From leninist to neoliberal biopolitics*. Palo Alto, CA: Stanford University Press.
Gries, P. N. (2005) *China's new nationalism: Pride, politics, and diplomacy*. Berkeley, CA: University of California Press.
Gross, E. .F., Juvonen, J., and Gable, S. L. (2002) Internet use and well-being in adolescence. *Journal of Social Issues* 58(1): 75–90.
The Guardian (November 11, 2008) *China recognizes Internet addiction as new disease*. At: http://www.guardian.co.uk/news/blog/2008/nov/11/china-internet (accessed August 5, 2008).
Guo, L. (2003). *The CASS Internet report 2003: Surveying Internet usage and impact in twelve Chinese cities*. Beijing: Research Center for Social Development, Chinese Academy for Social Sciences.
———. (2005) *Surveying Internet usage and impact in five Chinese cities*. Beijing: Chinese Academy of Social Sciences.
———. (2007) *Surveying Internet usage and impact in five Chinese cities*. Beijing Chinese Academy of Social Sciences.
Guthrie, D. (2006) *China and globalization: The social, economic and political transformation of Chinese society*. New York: Routledge.
Habermas, J. (1989) *The structural transformation of the public sphere*. Cambridge, UK: Polity Press.
Halbwachs, M. (1950) *The collective memory*. New York: Harper & Row.
Hall, C. M. (1990) *Women and identity: Value choices in a changing world*. New York: Hemisphere Publishing Corporation.
Hall, S. (1987) 'Minimal selves'. In *Identity: The real me*, L. Appignanesi, ed., 44–46. London: ICA.

———. (1988) New ethnicities. In *Black film/British cinema*, K. Mercer, ed., 27–31. London: ICA.

———. (1997) *Representation: Cultural representations and signifying practices*. London: Sage.

Hamrin, C. L. (1990) *China and the challenge of the future*. San Francisco: Westview Press.

Hardey, M. (2002) Life beyond the screen: embodiment and identity through the internet. *The Sociological Review* 50(4): 570–585.

Hardin, R. (1995) Self-Interest, group identity. In *Nationalism and Rationality*, Albert Breton, ed., 14–42. Cambridge, UK: Cambridge University Press.

Harrison, H. (2000) *The making of the republican citizen*. Oxford, UK: Oxford University Press.

He, B. G. and Guo, Y. J. (2000) *Nationalism, national identity and democratization in China*. Aldershot, UK: Ashgate Publishing.

Hechter, M. (2000) *Containing nationalism*. Oxford, UK: Oxford University Press.

Herring, S. C. (2010) Web content analysis: Expanding the paradigm. In *The international handbook of Internet research*, J. Hunsinger, M. Allen, and L. Klastrup, eds. Frankfurt: Springer.

Hetherington, K. (1997) *The badlands of modernity: Heterotopia and social ordering*. London: Routledge.

Hiller, H. H. and Franz, T. M. (2004) New ties, old ties and lost ties: The use of new media in diaspora. *New Media and Society* 6(6): 731–52.

Hine, C. M. (2000) *Virtual ethnography*. London: Sage.

Hobsbawm, E. (1983) Introduction: Inventing traditions and mass producing traditions: Europe, 1870–1914. In *The invention of traditions*, E. Hobsbawm and T. Ranger, eds., 1–14. Cambridge, UK: Cambridge University Press.

Hoffman, L. (2003) Enterprising cities and citizens: The re-figuring of urban spaces and the making of post-Mao professionals. *Provincial China* 8(1): 526.

———. (2006a) Autonomous choices and patriotic professionalism: On governmentality in late-socialist China. *Economy and society* 34(5): 550–570.

———. (2006b) Urban transformation and professionalism: Translocality and rationalities of enterprise in post-Mao China. In *Translocal China*, T. Oakes and L. Schein, eds., 109–137. London: RoutledgeCurzon.

Hong, J. and Huang, L. (2005) A split and swaying approach to building information society: The case of Internet cafés in China. *Telematics and Informatics* 22(4): 377–393.

Hoskins, B. L. (2005) Translating the research results on political participation into policy. In J. Forbrig, ed., *Revisiting youth political participation*. Strasbourg: Council of Europe Publishing.

Hsu, F. L. K. (1948) *Under the ancestors' shadow: Chinese culture and personality*. New York: Columbia University Press.

Hwang, K. K. (1999) Filial piety and loyalty: Two types of social identification in Confucianism. *Asian Journal of Social Psychology* 2: 163–183.

Hull, G. (2003) At last youth culture and digital media: New literacies for new times. *Research in the Teaching of English* 38(2): 229–33.

IAC and JWT (2007) Young digital mavens study, China's online youth lead U.S. counterparts in digital self-expression. At: http://www.marketingcharts.com/interactive/chinas-online-youth-lead-us-counterparts-in-digital-self-expression-2575/ (accessed May 28, 2009).

Ikels, C. (2006) Economic reform and intergenerational relationships in China. *Oxford Development Studies* 34(4): 387–400.

Jacob, W. J. (2006) Social justice in Chinese higher education: Regional issues of equity and access. *Review of Education* 52(1): 149–169.

James, R. (2009) A brief history of: Chinese Internet censorship. *Times*, March18, 2009. At: http://www.time.com/time/world/article/0,8599,1885961,00.html (accessed January 8, 2010).

Jameson, F. (1986) Third world literature in the era of multinational capitalism. *Social Text* (Fall 1986).

Jankowski, N. W. (2006) Creating community with media: History, theories and scientific investigations. In *The handbook of new media*, L. A. Lievrouw and S. Livingstone, eds., 55–74. London: Sage Publications.

Jasper, J. M. and Polletta, F. (2001) Collective identity and social movements. *Annual Review of Sociology* 27: 283–305.

Jennings, M. K. and Zeitner, V. (2003) Internet use and civic engagement: A longitudinal analysis. *Public Opinion Quarterly* 67: 311–334.

Jing, J. (2000) Introduction: Food, children, and social change in contemporary China. In *Feeding China's little emperors*, Jun Jing, ed., 1–26. Palo Alto, CA: Stanford University Press.

Johnson, A. (1997) Business ethics in China. *China Review* 7: 15–18.

Jones, G. and Wallace, C. (1992) *Youth, family and citizenship*. Open University Press.

Jones, S. G. (1995) Cybersociety: Computer-mediated communication and community. Thousand Oaks, CA: Sage.

Josselson, R. (1995). *Imagining the real. In Interpreting experience: the narrative study of lives*, R. Josselson and A. Lieblich, eds., 27–45. Thousand Oaks, Calif.: Sage Publications.

Josselson, R. (1996) On writing other people's lives. In *Ethics and process in the narrative study of lives*, R. Josselson, ed., 60–71. Thousand Oaks, CA: Sage Publications.

Jung, J. Y., Qiu, J. L. C. and Kim, Y. C. (2001) Internet connectedness and inequality: Beyond the 'digital divide'. *Communication Research* 28(2): 507–535.

Khan, A. R. and Riskin, C. (2000). *Inequality and poverty in China in the age of globalization*. Oxford, UK: Oxford University Press.

Kipnis, A. (1997) *Producing guanxi: sentiment, self and subculture in a north China village*. Durham, NC: Duke University Press.

———. (2006) Suzhi: A keyword approach. *The China quarterly* 186, 295–313.

Kluver, R. and Yang, C. (2005) The Internet in China: A meta-review of research. *The Information Society* 21(4): 301–308.

Kraut R, Patterson M, Lundmark V, Kiesler S, Mukopadhyay T, and Scherlis W. (1998) Internet paradox: a social technology that reduces social involvement and psychological well-Being? *American Psychologist* 53 (9): 1017–1031.

Kwong, J. (1994) Ideological crisis among China's youths: Values and official ideology. *British Journal of Sociology* 45(2): 247–264.

Lægran, A. S. (2002) The petrol station and the Internet café: Rural technospaces for youth. *Journal of Rural Studies* 18(2):157–168.

Lægran, A. S. and Stewart, J. (2003) Nerdy, trendy or healthy? Configuring the Internet café. *New Media and Society* 5(3): 357–377.

Lagerkvist, J. (2009) Contesting norms on China's Internet? The party-state, youth, and social change. In Youth engaging with the world: Media, communication and social change, T. Tufte and F. Enghel, eds., 215–231. Göteborg: Nordicom.

Lankshear, C. and Knobel, M. (2006) *New literacies: Everyday practices and classroom learning*. New York: Open University Press.

Lardy, N. (1983) *Agriculture in China's modern economic development*. Cambridge, UK: Cambridge University Press.

Latham, K. (2007) SMS, communication, and citizenship in China's information society, *Critical Asian Studies* 39(2): 295–314.

Lee, L. (2005) Young people and the Internet: From theory to practice. *Young* 13(4): 315–326.

Lee, S. (1999) Private uses in public spaces, a study of an Internet café. *New Media & Society* 1(3): 331–50.
Leu, D. J., Kinzer, C. K., Coiro, J. L. and Cammack, D. W. (2004) Toward a theory of new literacies emerging from the Internet and other information and communication technologies. In *Theoretical models and processes of reading*, R. B. Ruddell and N. Unrau, eds., 1570–1613. Newark, NJ: International Reading Association.
Li, N. and Kirkup, G. (2007) Gender and cultural differences in Internet use: A study of China and the UK. *Computers & Education* 48(2): 301–317.
Liff, S. and Lægran, A. S. (2003) Cybercafés: Debating the meaning and significance of Internet access in a Café environment. *New Media and Society* 5(3): 307–312.
Liff, S. and Steward, F. (2001) Community e-gateways: Locating networks and learning for social inclusion. *Information, Communication and Society* 4(3): 317–40.
———. (2003) Shaping e-Access in the Cybercafé: Networks, boundaries and heterotopian innovation. *New Media and Society* 5(3): 313–334.
Lim, S. S. (2008) Technology domestication in the Asian homestead: Comparing the experiences of middle class families in China and South Korea. *East Asian Science, Technology and Society: An International Journal* 2(2):189–209.
Lin, C. (2006) *The transformation of Chinese socialism*. Durham, NC: Duke University Press.
Lin, C. H. and Yu, S. F. (2008) Adolescent Internet usage in Taiwan: Exploring gender differences. *Adolescence* 43(170): 317–31.
Lin, J. (1999) *Social transformation and private education in China*. U.S.: Praeger.
Litzinger, R. (2002) Theorizing postsocialism: Reflections on the politics of marginality in contemporary China. *The South Atlantic Quarterly* 101(1): 33–55.
Liu, F. S. (2002a). *Parental expectations of the single child in the nuclear chinese family: Boys as single children and girls as single children*. M. Phil. thesis, Faculty of Education, University of Oslo.
———. (2002b) The dynamics of culture and the implications for education in China. *In Focus Journal*, 1 (Fall).
———. (2004) Basic education in China's rural areas: A legal obligation or an individual choice? *International Journal of Educational Development* 24(1): 5–21.
———. (2006a) *Modernization as lived experiences: Identity construction of young-adult only-children in present-day China*. PhD dissertation. Institute of Educational Research, University of Oslo, Norway.
———. (2006b) Boys as only-children and girls as only-children—Parental gendered expectations of the only-child in the nuclear Chinese family in present-day China. *Gender and Education* 18(5): 491–506.
———. (2006c) School culture and gender. In *The Sage handbook of gender and education*, C. Skelton, B. Francis and L. Smulyan, eds., 425–438. New York: Sage.
———. (2008a) Constructing the autonomous middle-class self in today's China: The case of young-adult only-children university students. *Journal of Youth Studies* 11(2): 193–212.
———. (2008b) Negotiating the filial self: Young-adult only-children and intergenerational relationships in China. *Young—Nordic Journal of Youth Research* 16(4): 409–430.
———. (2009) It is not merely about life on the screen: Urban Chinese youth and the Internet Café. *Journal of Youth Studies* 12(2): 167–184.
———. (2010) The Internet in the everyday lifeworld: A comparison between high school students in China and Norway. *Comparative Education* 46 (4): 527–550.
Liu, K. (2004) *Globalization and cultural trends in China*. Honolulu: University of Hawai'i Press.
Liu, S. D. (2005) China's popular nationalism on the Internet: Report on the 2005 anti-Japan network struggles. *Inter-Asia Cultural Studies* 7(1): 144–155.

Livingstone, S. (2002) *Young people and new media: Childhood and the changing media environment*. London: Sage.

———. (2003) Children's use of the Internet: Reflections on the emerging research agenda. *New Media & Society* 5: 147–66.

Livingstone, S. and Bober, M. (2004) Taking up online opportunities? Children's uses of the Internet for education, communication and participation. *E–Learning* 1(3): 395–18.

———. (2005) *UK children go Online: Final report of key project findings*. London: London School of Economics and Political Science. At: http://eprints.lse.ac.uk/399/1/UKCGO_Final_report.pdf (accessed December 10, 2005).

Livingstone, S., Bober, M. and Helsper, E. J. (2005) Active participation or just more information? *Information, Communication and Society* 8 (3): 287–314.

Livingstone, S., L. d'Haenens, and U. Hasebrink, U. (2001) Childhood in Europe: Contexts for comparison. In *Children and their changing media environment. A European comparative study,* S. Livingstone and M. Bovill, eds., 3–30. New Jersey: Lawrence Erlbaum Associates.

Lo, K. C. (2009) The web marriage game, the gendered self, and Chinese modernity. *Cultural Studies* 23(3): 381–403.

Lu, S. H. (2000) Global POSTmodernIZATION: The intellectual, the artist and China's condition. In *Postmodernism and China*, A. Dirlik and X. Zhang, eds., 145–174. Durham, NC: Duke University Press.

Lu, T. H. (2005) Zhongguo xiaofeizhe xingwei baogao [Chinese consumer behavior]. Beijing: zhongguo shehui kexue chubanshe.

Lum, T. (2006) *Internet development and information control in the People's Republic of China*. CRS report for Congress. Washington, D.C.: Library of Congress.

Ma, W. H. (2003) *Economic reform and higher education in China*. Center for International and Development Education (CIDE) Publications Series. Los Angeles, CA: CIDE.

Mackay, H. and Gillespie, G. (1992) Extending the social shaping of technology approach: Ideology and appropriation. *Social Studies of Science* (22)4: 685–71.

Mannheim, K. (1952/1923) The problem of generations. In *Essays on the sociology of knowledge*, Mannheim, K., ed., 276–320. London: Routledge and Kegan Paul.

Mansbridge, J. J. and Morris, A., eds. (2001) *Oppositional consciousness: The subjective roots of social protest*. Chicago, IL: University Of Chicago Press.

Markham, A. (1998) *Life Online: Researching real experience in virtual space.* Walnut Creek, CA: Altamira Press.

Marr, D. and Rosen, S. (1998) Chinese and Vietnamese youth in the 1990s. *The China Journal* 40: 145–172.

Mclaren, A. (2007) Online intimacy in a Chinese setting. *Asian Studies Review* 31(4): 409–422.

McMillan, Sally, J. and Morrison, Margaret (2006) Coming of age with the Internet: A qualitative exploration of how the Internet has become an integral part of young people's lives. *New Media & Society* 8(1): 73–95.

McNeal, J. U., and Ji, M. (1999) Children's influence on Chinese families' newfound leisure time and its marketing implications. *Asia Pacific Journal of Marketing and Logistics* 8: 32–57.

McNeal, J. U., and Wu, S. S. (1995) Chinese children influence 45 percent of grocery picks. *Market: Asia Pacific* 6: 3–4.

McNeal, J. U., and Yeh, C. H. (2003) Consumer behavior of Chinese children: 1995–2002. *Journal of Consumer Marketing* 20: 542–54.

Meisner, M. (1977) Harmony and conflict in the Maoist utopian vision. *Journal of Chinese Philosophy* 4: 247–259.

Meng, X. (2000). *Labour market reform in China*. Cambridge, UK: Cambridge University Press.
Meng, X., Gregory, R. and Wang, Y. J. (2005) Poverty, inequality, and growth in urban China, 1986–2000. *Journal of Comparative Economics* 33: 710–729.
Meyer, M., Michael, D. C. and Nettesheim, C. (2008) *China's digital generations: The 570-million-hour opportunity*. Boston: The Boston Consulting Group, Inc.
Miles, M. and Huberman, M. (1994) *Qualitative Data Analysis: An Expanded Sourcebook*. London: Sage.
Miller, D. and Slater, D. (2000) *The Internet: An ethnographic approach*. Oxford: Berg.
Ministry of Education (MoE) (2007) *Gross enrolment rate of schools by level*. Beijing: MoE.
Mitra, A. (2003) Marginal voices in cyberspace. *New Media and Society* 3(1): 29–48.
Mooney, P. (2004) 'Beijing's Internet dilemma', *South China Morning Post*, April 14, p. A13.
Moore, R. L. (2005) Generation ku: Individualism and China's millennial youth. *Ethnology* 44(4): 357–376.
Mosco, V. (2004) *The digital sublime: Myth, power, and cyberspace*. Cambridge, MA: MIT Press.
Nelson, L. J. and Chen, X. Y. (2007) Emerging adulthood in China: The role of social and cultural factors. *Child Development Perspective* 1(2): 86–91.
Newcomb, T. M. (1950) *Social psychology*. New York: Dryden Press.
Ngai, N. P., Cheung, C. K. and Li, C. K. (2001) China's youth policy formulation and youth participation. *Children and Youth Services Review* 23(8): 651–669.
Nie, N. . H. and Erbring, L. (2000). *Internet and society: A preliminary report*. Palo Alto, CA: Stanford Institute for the Quantitative Study of Society.
Niko Partners (2010) *Annual review and five-Year forecast report on China's video game industry*. At: http://www.nikopartners.com/pdf/NikoPressRelease033110.pdf (accessed May 15, 2010).
Nilan, P. and Feixa, C. (2006) Introduction: Youth hybridity and plural worlds. In *Global youth? Hybrid identities, plural worlds*, P. Nilan and C. Feixa, eds., 1–13. London: Routledge.
Nip, J. Y. M. (2004) The queer sisters and its electronic bulletin board. *Information, Communication & Society* 7(1): 23–49.
Nolan, P. and Paine, S. (1987). Towards an appraisal of the impact of rural reform in China, 1978–1985. In *The re-emergence of Chinese peasantry*, A. Saith, ed., 81–104. London: Croom Helm.
Nolan, P. and White, G. (1984) Urban bias, rural bias or state bias? Urban-rural relations in post-revolutionary China. *Journal of Development Studies* 20: 52–81.
Norris, P. (2002) *Democratic phoenix: Reinventing political activism*. Cambridge, UK: Cambridge University Press.
———. (2005) The impact of the Internet on political activism. *International Journal of Electronic Government Research* 1(1): 19–39.
O'Toole, T. (2003) Engaging with young people's conceptions of the political. *Children's Geographies* 1(1): 71–90.
Oblinger, D. G. and Oblinger, J. L. (2005) *Educating the Net generation*. Boulder, Colorado: EDUCAUSE.
Oldenburg, R. (1999) *The great good place: Cafés, coffee shops, bookstores, bars, hair salons and other hangouts at the heart of a community*. New York: Marlowe and Company.

Ong, A. (2003) Cyberpublics and diaspora politics among transnational Chinese. *Interventions* 5(1): 82–100.
OpenNet Initiative (2005) *Internet Filtering in China in 2004–2005: A country study*. At: http://www.opennetinitiative.net/studies/china (accessed September 2009).
Özkirimli, U. (2000) *Theories of nationalism: A critical introduction*. London: Macmillan.
Palmer, M. (1995) The re-emergence of family law in post-Mao China: Marriage, divorce and reproduction. *The China Quarterly* 141: 110–134.
Panagakos, A. (2003) Downloading new identities: Ethnicity, technology and media in the global Greek village. *Identities* 10(2): 201–19.
Parham, A. A. (2004) Diaspora, community and communication: Internet use in transnational Haiti. *Global Networks* 4(2): 199–217.
Parker, D. and Song, M. (2006) Ethnicity, social capital and the Internet: British Chinese websites. *Ethnicities* 6(2): 178–202.
PC Magazine (2006) A dangerous Net (November 21, 2006), 23.
Peng S. Q. (2003) *Wo Ping shenmo xinren ni?* [By what Should I trust you?] In Zheng Yifu and Peng Siqing, eds., *Zhongguo shehui zhong de xinren* [Trust in Chinese society]. Beijing: Zhongguo chengshi chubanshi, 292–301.
People's Daily Online (August 2, 2000) *Chinese leaders' remarks on youth work published*. At: http://english.peopledaily.com.cn/english/200008/02/eng20000802_47136.html (accessed April 20, 2010).
———. (June 3, 2003) *China's first generation of the only-child to step into marriage 'citadel'*. At: http://english.people.com.cn/200306/03/eng20030603_117623.shtml (accessed June 3, 2005).
———. (October 27, 2004) *Dissecting China's 'middle class'*. At: http://english.peopledaily.com.cn (accessed November 12, 2005).
———. (March 7, 2005), *Ph.D.s queue up for high school teaching jobs*. http://english.people.com.cn/200503/07/eng20050307_175927.html (accessed December 12, 2005).
———. (October 5, 2006) *Teenagers' buying power expands*. At: http://english.peopledaily.com.cn/200610/05/eng20061005_309197.html (accessed April 3, 2010).
Perkins, D. and Yusuf, S. (1984) *Rural development in China*. Baltimore, MD: John's Hopkins University Press.
Pew Research Center (2005) *China's optimism: Prosperity brings satisfaction—and hope. Summary of findings*. Pew Global Attitudes Project. At: http://pewglobal.org/reports/display.php?ReportID=249 (accessed May 1, 2010).
Poon, A. (2008) Chinese youth need own group belief. *Asia Sentinel*. At: http://www.asiasentinel.com/index.php?option=com_content&task=view&id=1188&Itemid=158 (accessed March 3, 2009).
Potter, J. and Wetherell, M. (1987) Discourse and social psychology: Beyond attitudes and behaviour. London: Sage.
———. (1995) Discourse analysis. In *Rethinking methods in psychology*, J. A. Smith, R. Harré and L. Van Langenhove, eds., 80–92. London: Sage,.
Prensky, M. (2001) Digital natives and digital immigrants: A new way to look at ourselves and our kids. *On the Horizon* 9(5): 1–6.
———. (2006) *Don't bother me, mom—I'm learning!* St. Paul, MN: Paragon House.
Putnam, R. D. (1995) Bowling alone: America's declining social capital. *Journal of Democracy* 6(1): 65–78.
———. (2000) *Bowling alone: The collapse and revival of American community*. New York: Simon and Schuster.
Pye, L. (1990) China: Erratic state, frustrated society. *Foreign Affairs* 69(4): 56–74.

Qiang, C. Z. W. (2007) *China's information revolution: managing the economic and social transformation.* Washington, D.C.: World Bank.

Qiu, J. L. C. (2003) *The Internet in China: Data and issues.* Working paper prepared for Annenberg research seminar on international communication. At: http://arnic.info/Papers/JQ_China_and_Internet.pdf (accessed May 2, 2004).

Qiu, J. L. C. and Zhou, L. N. (2005) Through the prism of the Internet café: Managing access in an ecology of games. *China Information* 19: 261–97.

Rawski, T. (1979) *Economic growth and employment in China.* Oxford: Oxford University Press.

Research and Market Report (2007) *China advertising industry report, 2006–2007.* At: http://www.researchandmarkets.com/reports/451693 (accessed March 4, 2010).

Rheingold, H. (2000) *The virtual community: Homesteading on the electronic frontier.* Revised edition. Cambridge, MA: MIT Press.

Rice, R. and Katz, J. (2004) The Internet and political involvement in 1996 and 2000. In *Society online: The Internet in context*, P. Howard and S. Jones, eds., 103–120. Thousand Oaks, CA.: Sage Publications.

Rich, L. and Tsui, M. (2002) The only child and educational opportunities for girls in urban China. *Gender and Society* 16(1): 74–92.

Riesman, D. (with Glazer, N. and D. Reuel) (1989/1961) *The lonely crowd: A study of the changing American character.* New Haven, CT: Yale University Press.

Riskin, C. (1987) *China's political economy: The quest for Development since 1949.* Oxford: Oxford University Press.

Riskin, C., Zhao, R. and Li, S. (2001) *China's retreat from equality: Income distribution and economic transition.* Armonk, NY: Me. E. Sharpe.

Rose, N. (1996a) Authority and the genealogy of the subject. In *Detraditionalization: Critical reflections on authority and identity*, P. Heelas, S, Lash, and P. Morris, eds., 294–327. Malden, MA: Blackwell.

———. (1996b) Identity, genealogy, history. In *Questions of cultural identity*, S. Hall and P. Du Gay, eds., 128–150. London: Sage Publications.

———. (1998) *Inventing ourselves.* Cambridge, UK: Cambridge University Press.

Rosen, S. (2004) The state of youth/youth and the state in early 21st century China: The triumph of the urban rich? In *State and society in 21st century China: Crisis, contention, and legitimation*, P. H. Gries and S. Rosen eds., 159–179. London: Routledge.

———. (2009) Contemporary Chinese youth and the state. *The Journal of Asian Studies* 68 (2): 359–369.

Rosenberg, B. G., and Jing, Q. C. (1996) A revolution in family life: The political and social structural impact of China's one child policy. *Journal of Social Issues* 52: 51–79.

Sande, G. N., Goethals, G. R., and Radloff, C. E. (1988) Perceiving one's own traits and others': The multifaceted self. *Journal of Personality and Social Psychology* 54: 13–20.

Sandvig, C. (2006) The Internet at play: Child users of public internet connections. *Journal of Computer-Mediated Communication* 11(4): 932–956.

Sautedé, E. (2003) The snares of modernity: Internet, information and the SARS crisis in China. *China Perspectives*, May–June. At: http://chinaperspectives.revues.org/document273.html (accessed May 4, 2007).

Schutz, A. (1970) Reflections on the problem of relevance. Edited by R. M. Zaner. New Haven, CT: Yale University Press.

Schutz, A. and Luckmann, T. (1973) *The structures of the life-world.* Evanston, IL: North-Western University Press.

Sedikides, C. and Wildschut, T. (2008) Nostalgia as enabler of self-continuity. In *Individual and collective self-continuity: Psychological perspectives*, F. Sani, ed., 227–242. Mahwah, NJ: Lawrence Erlbaum Associates.

Sedikides, C., Wildschut, T. and Baden, D. (2004). Nostalgia: Conceptual issues and existential functions. In *Handbook of experimental existential psychology*, J. Greenberg, S. Koole and T. Pyszczynski, eds., 200–214. New York: Guilford Press.
Sefton-Green, J. (2004) *Literature review in informal learning with technology outside school*. Bristol, NESTA Futurelab. At: http://www.futurelab.org.uk/resources/documents/lit_reviews/Informal_Learning_Review.pdf (accessed March 15, 2005).
Seiter, E. (2005) *The Internet playground: Children's access, entertainment, and mis-education*. New York: Peter Lang.
Selber, S. A. (2004) *Multiliteracies for a digital age*. Carbondale: Southern Illinois University Press.
Shah, D., Cho, J., Eveland, W. P. and Kwak, N. (2005) Information and expression in a digital age: Modelling Internet effects on civic participation. *Communication Research* 32 (5): 531–565.
Shirk, S. L. (2007) *China: Fragile superpower*. Oxford: Oxford University Press.
Shotter, J. (1984) *Social accountability and selfhood*. Oxford: Blackwell.
Sibley, D. (1995) *Geographies of exclusion*. London: Routledge.
Sigley, G. (2004) Liberal despotism: Population planning, subjectivity and government in contemporary China. *Alternatives* 29(5): 557–575.
———. (2006) Chinese governmentalities: Government, governance and the socialist market economy. *Economy and Society* 35(4): 487–508.
Skelton T. (2000) 'Nothing to do, nowhere to go?' Teenage girls and 'public' space in the Rhondda Valleys, South Wales. In *Childrens' geographies*, S. L. Holloway and G. Valentine, eds., 69–85. London: Routledge, London.
Skelton, T., and Valentine, G., eds. (1998) *Cool places: Geographies of youth cultures*. London: Routledge.
Slater, D. (2001) Social relationships and Identity online and offline. In *The handbook of new media*, L. A. Lievrouw and S. Livingstone, eds., 533–46. London: Sage.
Snow, D. A. and McAdam D. (2000) Identity work processes in the context of social movements: Clarifying the identity/movement nexus. In *Self, identity and social movements*, S. Stryker, T. Owens T., and R. W. White, eds., 533–46. Minneapolis, MN: University of Minnesota Press.
Song, S. F. (2003) Policy issues of China's urban unemployment. *Contemporary Economic Policy* 21(2): 258–269.
Stanat, M. (2006) China's generation Y: *Understanding the future leaders of the world's next superpower*. Paramus, NJ.: Homa & Sekey Books.
Stapleton, J. (2005) Citizenship versus patriotism in twentieth-century England. *The Historical Journal* 48(1): 151–178.
Stefik, M. (1996) *Internet dreams: Archetypes, myths, and metaphors*. Cambridge, MA: MIT Press.
Stewart, J. (2000) Cafematics: cybercafés and the community. In *Community informatics: Enabling communities with information and communication technologies*, M. Gurstein, ed., 320–338. Hershey, PN: Idea Group Publishing.
St-Maurice, I. and Wu, C. (2006) Understanding China's teen consumers. *The McKinsey Quarterly*. At: http://www.mckinseyquarterly.com/Understanding_Chinas_teen_ consumers_1797 (accessed March 10, 2010).
Stockman, N. (2000) *Understanding Chinese society*. Cambridge, UK: Polity Press.
Stolle, D., Hooghe, M. and Micheletti, M. (2005) Politics in the supermarket: Political consumerism as a form of political participation. *International Political Science Review* 26(3): 245–269.

Stone, A. R. (1991) Will the real body please stand up: Boundary stories about virtual cultures. In *Cyberspace: First steps*, M. Benedikt, ed., 81–118. Cambridge, MA: MIT Press.

The Straits Times (January 16, 2009) *China's income gap wider*. At: http://www.straitstimes.com/Breaking%2BNews/Money/Story/STIStory_326966.htm (accessed January 4, 2010).

Street, B. (2003) What's "new" in New Literacy Studies? Critical approaches to literacy in theory and practice. *Current Issues in Comparative Education* 5(2): 77–91.

Sun, H. Y. (2003) The current status of Chinese children. *Journal of Family and Economic Issues* 24: 337–53.

Tai, Z. X. (2006) *The Internet in China: Cyberspace and civil society*. New York: Routledge.

Tan, Z. X. (1999) Regulating China's Internet: Convergence toward a coherent regulatory regime. *Telecommunications Policy*, 23(3), 261–276.

Tang, X. B. (1996) *Global space and the national discourse of modernity: The historical thinking of Liang Qichao*. Palo Alto, CA: Stanford University Press.

Tapscott, D. (1998) *Growing up digital: The rise of the Net generation*. New York: McGraw-Hill.

———. (1999) Educating the Net generation, *Educational Leadership* 56(5): 6–11.

Taubman, G. (1998) A not-so world wide web: The Internet, China, and the challenges to nondemocratic rule. *Political Communication* 15(2): 255–272.

Tian, B. X. (2005). *Zhang Chunliang: Damned Internet Games*. At http://www.southcn.com/nfsq/ywhc/tbxst/shentan/200510260380.htm (accessed January 20, 2007).

Tian, C. S. (2005) Chinese dialectics: from Yijing to Marxism. Lanham: Lexington Books.

Timesonline (June 7, 2009) *China's parents try shock tactics to cure Net 'addicts'*. At: http://www.timesonline.co.uk/tol/news/world/asia/china/article6445982.ece (accessed December 4, 2009).

Townsend, J. (1996) Chinese nationalism. In *Chinese nationalism*, J. Unger, ed., 1–30. Armonk, NY: M. E. Sharpe.

Tsui, L. (2005) Introduction: The socio-political Internet in China, *China Information* 19(2): 181–88.

———. (2008) *The great firewall as Iron Curtain 2.0: The implications of China's Internet most dominant metaphor for U.S. Foreign Policy*. Paper for the 6th annual Chinese Internet research conference. June 13–14, 2008. Hosted by the Journalism and Media Studies Centre, Hong Kong University.

Turkle, S. (1984) *The second self: Computers and the human spirit*. New York: Simon & Schuster.

———. (1995) *Life on the screen: Identity in the age of the Internet*. New York: Touchstone Books.

Uotinen, J. (2003) Involvement in (the information society): The Joensuu Community Resource Centre NetCafé. *New Media & Society* 5(3): 335–56.

Uslaner, E. (2004) Trust, civic engagement, and the Internet. *Political Communication* 21(2): 223–242.

Verba, S. and Nie, N. (1972) Participation *in America-political democracy and social equality*. New York: Harper & Row.

Wakeford, N. (1999) Gender in the landscape of computing. In *Virtual geographies*, M. Crang, P. Crang and J. May, eds., 178–201. London: Routledge.

———. (2003) The embedding of local culture in global communication: Independent Internet cafes in London, *New Media & Society* 5: 379–399.

Wallerstein, I. (1998) *Utopistics, or, historical choices of the twenty-first century*. New York: New Press.

Walton, G. (2001) *China's golden shield*. International Centre for Human Rights and Democratic Development. At: http://www.ichrdd.ca/english/commdoc/publications/globalization/goldenMenu.html (accessed January 10, 2007).
Wang, J. (2008) *Brand new China: Advertising, media, and commercial culture*. Cambridge, MA: Harvard University Press.
Wang, J. S. (2001) The Internet in China: A new fantasy? *New Perspectives Quarterly* 18(1): 1–3.
Wang, S. B. and Zhai, B. (1999) *Education through suffering: the Bible for Chinese parents [Kunan Jiayu—Zhongguo Fumu Yuer Shengjin]*. Xinan: Xinan Teachers' University Press.
Wang, S. S. and Hong, J. H. (2010) Discourse behind the forbidden realm: Internet surveilance and its implications on China's blogosphere. *Telematics and Informatics* 27: 67–78.
Wang, Y. (2007) Kids forced to work too hard: Survey. *China Daily*. http://www.chinadaily.com.cn (Accessed June 5, 2010).
Washington Post (February 21, 2007) Strict clinic near Beijing is operated by a military researcher. At: http://www.washingtonpost.com/wpdyn/content/article/2007/02/21/AR2007022102094.html (accessed February 24, 2010).
Weber, I. (2002) Shanghai baby: Negotiating youth self-identity in urban China. *Social Identities* 8(2): 347–368.
———. (2003) Localizing the global: Successful strategies for selling television programmes to China, *Gazette* 65(3): 273–90.
Weber, I. and Jia, L. (2007) Internet and self-regulation in China: the cultural logic of controlled commodification. *Media, Culture and Society* 29(5): 772–89.
Wellman, B. and Gulia, M. (1999) Net surfers don't ride alone. In *Networks in the global village*, B. Wellman, ed., 72–86. Boulder, CO: Westview.
Wertsch, J. V. (2002) *Voices of collective remembering*. Cambridge, UK: Cambridge University Press.
West, L. A. and Zhao Y. (2000) *Rural labor flows in China*. Berkeley, CA: Institute of East Asian Studies, University of California.
Wetherell, M. (1998) Positioning and interpretative repertoires: Conversation analysis and post-structuralism in dialogue. *Discourse and Society* 9(3): 387–412.
White, C., Bruce, S. and Ritchie, J. (2000) *Young people's politics: Political interest and engagement amongst 14–24 Year Olds*. York: Joseph Rowntree Foundation.
White, G. (1996) Corruption and the transition from socialism in China. *Journal of Law and Society* 23: 149–64.
Whyte, M. K., ed. (1994) *China's revolutions and intergenerational relations*. Ann Arbor, MI: Center for Chinese Studies, University of Michigan.
Wilhelm, A. G. (2000) *Democracy in the digital age: Challenges to political life in cyberspace*. New York: Routledge.
Willis, P. (1990) *Common culture*. Boulder, CO: Westview.
World Bank (1997) *Sharing rising incomes: Disparities in China*. Washington, D.C.: The World Bank.
———. (2009) *From poor areas to poor people: China's evolving poverty reduction agenda. An assessment of poverty and inequality in China*. Washington, D.C.: Poverty Reduction and Economic Management Department. East Asia And Pacific Region, World Bank.
World Trade Organization (WTO) (2008) *International trade statistics 2008*. Geneva: WTO. At: http://www.wto.org/english/res_e/statis_e/its2008_e/its08_toc_e.htm (accessed March 5, 2010).
Woronov, T. (2002) *Transforming the future: 'Quality' children for the Chinese nation*. PhD. Dissertation, University of Chicago.

Wu, J. L. (2005) *Understanding and interpreting Chinese economic reform*. Mason, Ohio: Thomson/South-Western.

Wu, W., Fore, S., Wang, X. and Ho, P. S. Y. (2007) Beyond virtual carnival and masquerade: In-game marriage on the Chinese Internet. *Games and Culture* 2(1): 59–89.

Wu, X. (2007) *Chinese cyber nationalism: Evolution, characteristics and implications*. Lanham, MD: Lexington Books.

Wulff, H. (1995) Introduction: Introducing youth culture in its own right: the state of art and new possibilities. In *Youth cultures: A Cross Cultural Perspective*, V. Amit-Talai and H. Wulff, eds., 1–18. London: Routledge.

Xie, B. and Jaeger, P. (2008) Older adults and political participation on the Internet: A cross-cultural comparison of the United States and China. *Journal of Cross-Cultural Gerontology* 23(1): 1–15.

Xinhua Daily (April 21, 2008) *Patriotism requires passion, but more importantly, it requires rationality*. [aiguo xuyao jiqing, geng xuyao lixing]. http://news.xinhuanet.com/politics/2008–04/21/content_8019925.htm (accessed March 5, 2010).

Xinhua News Agency (December 29, 2007) *'Internet love' big for university students*. At http://www.china.org.cn/english/China/237453.htm (accessed April 6, 2008).

Xu, B. (1999) *Disenchanted democracy: Chinese cultural criticism after 1989*. Ann Arbor, MI: The University of Michigan Press.

———. (2001) Chinese populist nationalism: Its intellectual politics and moral dilemma. *Representations* 76: 120–140.

Xu, L. (2002) *Searching for life's meaning: Changes and tensions in the worldviews of Chinese youth in the 1980s*. Ann Arbor, MI: University of Michigan Press.

Yan, Y. X. (1996) *The flow of gifts: reciprocity and social networks in a Chinese viallage*. Palo Alto, CA: Stanford University Press.

———. (2005) Of hamburger and social space: Consuming McDonald's in Beijing. In *The cultural politics of food and eating: A reader*, J. L. Watson and M. L. Caldwell, eds., 80–103. Malden, MA: Blackwell.

———. (2006) Little emperors or frail pragmatists? China's '80ers generation. *Current History* 105(692): 255–62.

———. (2009) *The Individualization of Chinese society*. Oxford: Berg.

Yang, D. T. (1999) Urban-biased policies and rising income inequality in China. *American Economic Review* 89(2): 306–310.

Yang, G. B. (2003a) The Internet and civil society in China: A preliminary assessment, *Journal of Contemporary China* 12(36): 453–75.

———. (2003b) China's zhiqing generation: Nostalgia, identity, and cultural resistance in the 1990s. *Modern China* 29(3): 267–296.

———. (2009a) The Internet as cultural form: Technology and the human condition in China. *Know Techn Pol* 22: 109–115.

———. (2009b) *The Power of the Internet in China: Citizen activism online*. New York: Columbia University Press.

Yang, M. M. H. (1994) *Gifts, favors and banquets: the art of social relationships in China*. Ithaca, NY: Cornell University Press.

Yin, Q. P. and White, G. (1994) The "marketisation" of Chinese higher education: A critical assessment. *Comparative Education* 30(3): 217–237.

Ying, G. (2003) Consumption patterns of Chinese children. *Journal of Family and Economic Issues* 24 (4): 373–379.

You, J. (1998) *China's enterprise: Changing state/society relations after Mao*. London: Routledge.

Yu, H. Q. (2007a) Blogging everyday life in Chinese Internet culture. *Asian Studies Review*, 31(4): 423–433.
———. (2007b) Talking, linking, clicking: The politics of AIDS and SARS in urban China. *Positions: east asia cultures critique* 15(1): 35–63.
———. (2009) *Media and cultural transformation in China*. New York: Routledge.
Yueh, L. (2006) Parental investment in children's human capital in urban China. *Applied Economics* 38: 2089–2111.
Zhan, H. J. and Montgomery, R. J. V. (2003) Gender and elder care in China: The influence of filial piety and structural constraints. *Gender and Society* 17(2): 209–229.
Zhang, J. H. (2002) Will the government 'serve the people'? The development of Chinese E-government, *New Media & Society* 4(2): 163–184.
Zhang, T. J. (2004) Self-identity construction of the present China, *Comparative Strategy* 23(3): 281–301.
Zhang, W. (2008) The war between two labels: A true record of the 'post-80's' and 'post-90's' attacks, *China Youth Daily*, May 7, p.10.
Zhang, W. D. (2002) Displaying culture, voicing identity: A study of the Manchurian Website "Eight banners descendant." *NETCOM: Networks and Communications Studies* 16(3–4): 175–190.
Zhang, X. (2008) Ziyou (Freedom), Occupational choice, and labor: Bangbang in Chongqing, People's Republic of China. *International Labor and Working-Class History* 73(Spring): 65–84.
Zhao, D. X. (2002) Problems of nationalism in contemporary China: Student-government conflicts during nationalistic protests. In *Exploring nationalisms of China: Themes and conflict*, C. X. George Wei and X. Y. Liu, eds., 101–118. Westport, CT: Greenwood.
Zhao, R. W. (1993) Three features of the distinctions of income during the transition to reform. In *The distribution of income in China*, K. Griffin & R. W., Zhao, eds., 74–92. Basingstoke, UK: Macmillan.
Zhao, S. S. (2004) *Nation-state by construction: Dynamics of modern Chinese nationalism*. Palo Alto, CA: Stanford University Press.
———. (2005) China's pragmatic nationalism: Is it manageable? *The Washington Quarterly* 29(1): 131–144.
Zhao, Y. Z. (2007) After mobile phones, what? Re-embedding the social in China's "digital revolution." *International Journal of Communication* 1: 92–120.
Zheng Y. N. (1999) *Discovering Chinese nationalism in China: Modernization, identity, and international relations*. Cambridge, UK: Cambridge University Press.
Zhong, Y. (2002) Debating with muzzled mouth: A case analysis of how control works in a Chinese television debate used for educated youths, *Media, Culture & Society* 24(1): 27–47.
Zhou, M. C. and Lin, J. (2009) Chinese graduates' employment: The impact of the financial crisis, *International Higher Education* 55 (Spring). At: http://www.bc.edu/bc_org/avp/soe/cihe/newsletter/Number55/p3_Zhou&Lin.htm (accessed February 8, 2010).
Zhou, Y. M. (2006) *Historicizing online politics: Telegraphy, the Internet, and political participation in China*. Palo Alto, CA: Stanford University Press.
Zittrain, J. and Edelman, B. (2002) *Empirical analysis of Internet filtering in China*. Berkeman Center for Internet and Society, Harvard Law School. At: http://cyber.law.harvard.edu/filtering/china/ (accessed March 5, 2009).

Index

Note: Page numbers ending in 'n' refer to notes.

For Product Safety Concerns and Information please contact our EU representative GPSR@taylorandfrancis.com
Taylor & Francis Verlag GmbH, Kaufingerstraße 24, 80331 München, Germany

www.ingramcontent.com/pod-product-compliance
Ingram Content Group UK Ltd.
Pitfield, Milton Keynes, MK11 3LW, UK
UKHW022118190926
475702UK00013B/1511
* 9 7 8 0 4 1 5 8 5 1 9 1 6 *